Advancing Global Bioethics

Volume 23

The book series Global Bioethics provides a forum for normative analysis of a vast range of important new issues in bioethics from a truly global perspective and with a cross-cultural approach. The issues covered by the series include among other things sponsorship of research and education, scientific misconduct and research integrity, exploitation of research participants in resource-poor settings, brain drain and migration of healthcare workers, organ trafficking and transplant tourism, indigenous medicine, biodiversity, commodification of human tissue, benefit sharing, bio-industry and food, malnutrition and hunger, human rights, and climate change.

Beatrice Okyere-Manu • Peter Mwipikeni
Editors

An African Approach to Population Ethics

Issues, Questions, and Viewpoints

Springer

Editors
Beatrice Okyere-Manu
Department of Philosophy and
Ethics Studies
University of KwaZulu-Natal
Pietermaritzburg, KwaZulu-Natal
South Africa

Peter Mwipikeni
Department of Philosophy and
Ethics Studies
University of KwaZulu-Natal
Pietermaritzburg, KwaZulu-Natal
South Africa

ISSN 2212-652X ISSN 2212-6538 (electronic)
Advancing Global Bioethics
ISBN 978-3-032-24463-5 ISBN 978-3-032-24464-2 (eBook)
https://doi.org/10.1007/978-3-032-24464-2

This Springer imprint is published by the registered company Springer Nature Switzerland AG
The registered company address is: Gewerbestrasse 11, 6330 Cham, Switzerland

Foreword

This collection of essays is a welcome addition on African ethics as it deals with a very crucial topic. The issue of population is very important as it touches on a number of existential matters. Just as overpopulation is worrying, so is underpopulation—for different reasons and possible consequences. The situation in Africa is perilous on a number of fronts, and population increase adds to the fragility of Africa. While some arguments have been made about Africa's potential from its human resource potential, this is counterweighed by Africa's standing as the poorest place on earth. Poverty and a large population are not an ideal combination. This is especially the case when the majority of the population is young people or youths. The greatest contradiction of our time is the dissonance between Africa's natural wealth and human resources, on the one hand, and a distinct lack of prospect for millions of individuals. The search for prospects has led to many internal dynamics, such as rural to urban migration, leading to overpopulated cities. It is also seen in what others have called internal migration within the continent, which leads to flare-ups such as xenophobia. These existential matters must be taken up by thinkers committed to the good of the continent. Ethicists are one such group. Yet, when coming to thinking of what is good for the continent or what the good of the continent is, ethicists have been trapped by systems of thought that have often only spoken to one aspect or reality of Africa. The trap has been the attempt to make Africa's traditions the foundational view of all African reality. Traditional Africa was just but one episode. It was an episode that did not face the problems we face today, such as overpopulation and its resultant spin-offs, such as migration. In looking at Africa's modern problems through the traditional lens, whether dressed up as traditional ethics or as authentic African modes of thinking, we should ask ourselves how far such modes are geared to face up to today's problems. While it is important to be grounded in some ethical foundation that speaks to timeless ethical practices, there is a need to think of Africa as an evolving reality that demands the generation of new ideas that address the reality of the here and now. As the editors of this book correctly point out, the absence of an African view on population ethics only serves to deprive the field of the diversity that Africans could add. However, that addition must not be the sake of addition or even appropriation. It must firstly address real issues and secondly further the debate by generating new ideas. In part, this is what the volume seeks to achieve. While it is not possible to banish traditionalist-based views of African reality, the book has managed to add to its contributions a few entries that seek to tackle the real problem of overpopulation on the continent. These entries, hopefully, will contribute to the initiation of new and interesting debates on issues such as individual responsibility for procreation vis-à-vis moral and ontological categories of new and older theories. In particular, one thing that Africa has lagged in is planning. This is evidenced by the chaotic nature of governance, the lack of accountability of public entities, badly modeled budgetary plans that fail to adequately cover the needs of the masses, and the non-existence of coherent population plans. The growth of populations has a real bearing on the good of a country and the people in it. The expansion of numbers in Africa needs a plan that directly addresses the future fortunes of those being born. Those fortunes are, in a major way, dependent on the current fortunes of this generation. The fortunes of this generation will enable their planning for a favorable experience of life for future generations. Such planning must be backed by real and tangible reserves of wealth that are held in common. Yet most African countries run on unbelievable deficits that cripple both current and future generations. This perilous state of affairs surely must invite serious ethical questions on who bears responsibility for bringing life into existence and what they can legitimately expect from dysfunctional governments in the project of securing the future of present generations and those yet to come.

University of Pretoria
Pretoria, South Africa

Bernard Matolino

Acknowledgments

This book is the result of a collaborative effort, dialogue, and commitment from a diverse group of scholars who have critically engaged with issues of population ethics from African contexts and perspectives.

We would like to thank all the contributors whose chapters have enriched this volume with insightful analyses, thought-provoking questions, and grounded reflections. Your hard work and dedication have made this publication possible.

We also acknowledge the invaluable editorial support we received from our reviewers and editors at Springer. We extend our heartfelt thanks to Prof. Bernard Matolino, who graciously wrote the foreword of this book

Competing Interests The authors have no competing interests to declare that are relevant to the content of this manuscript.

Contents

Part I Foundations of African Population Ethics

1 **Introductory Chapter: Giving a Voice to African Ethics and Views in the Discourse on Population Ethics** 3
Beatrice Okyere-Manu and Peter Mwipikeni

2 **African Ethics and Population Issues** 13
Wilfred Lajul

3 **Rethinking Traditional Akan Proverbs on Fertility in the Context of Modern Population Challenges** 27
Beatrice Okyere-Manu

Part II Cultural and Philosophical Perspectives

4 **Meaning of Life and Population Issues from an African Perspective** 45
Wilfred Lajul

5 **Tension Between Procreation as a Duty and Individual Right in African Population Ethics** 59
Nomatter Sande

6 **Rethinking Procreation in the Era of Climate Crisis: An Ethico-Cultural Critique** 79
Hupenyu Makusha

Part III Challenges and Implications

7 **Necroeconomic Population Policy, the Racialized Capitalist World and the Surplus Population of Disposable Beings** 101
Peter Mwipikeni

8 **Rethinking Population Ethics and Environmental Sustainability through African Communitarian Thought** 123
Stephen Nkansah Morgan

9 **An African Population Ethics and the Environment: Addressing the Impact of Global Warming on the Well-Being of African Peoples** 141
Keba Cyril Muko

10 **Population Boom Versus Scarcity of Resources in Malawi: An African Historical-Ethical Critique** 171
Pascal Newbourne Mwale and Cobbener W. Sungani

Part IV Case Studies and Contemporary Issues

11 **An Ethical Exploration of Young People's Perception of Human Population Growth in Contemporary South Africa** 193
Nothando Happy-Girl Shandu and Beatrice Okyere-Manu

12 **Global Capitalism, Overpopulation in Africa, and African Population Ethics** 207
Peter Mwipikeni

13 **Xenophobia and Migration in South Africa: An Afro-Communitarian Approach to Population Ethics** 229
Lemohang Tebeli

About the Editors and Contributors

About the Editors

Peter Mwipikeni is a Lecturer in Ethics and Philosophy at the University of KwaZulu Natal. Peter teaches ethical theories, African philosophy, bioethics, leadership ethics, African ethics, and applied ethics. At present, his research focus is on African philosophy, political philosophy, philosophy of law, applied ethics, philosophy of religion, political economy, climate justice, AI, and ethics and decolonization. Before joining the University of KwaZulu-Natal in 2023, Peter was a lecturer at the University of Zululand.

Beatrice Okyere-Manu is an Associate Professor at the School of Religion, Philosophy, and Classics at the University of KwaZulu-Natal. She is the Programme Director for Applied Ethics. Her research interests cover environmental ethics, ethics of technology and AI, and African ethics. She brings together practical experience in community involvement with theoretical scholarship in African ethics. She has written several journal articles and book chapters. Her edited and co-edited books include *Intersecting African Indigenous Knowledge Systems and Western Knowledge Systems: Moral Convergence and Divergence* (2018); *African Values, Ethics, and Technology: Questions, Issues, and Approaches* (2021), *Development Ethics: an African Perspective* (2023), *African Women's Liberating Philosophies, Theologies, and Ethics* (2024), and *The Palgrave Handbook of Ubuntu, Inequality and Sustainable Development* (2024).

About the Contributors

Wilfred Lajul is a Professor of Philosophy and the current Head of the Religious Studies and Philosophy Department at Gulu University. Wilfred has two undergraduate degrees in philosophy and Theology. He has an MA in Philosophy (1992) and a PhD in Philosophy (1994) from Urban University, Rome. He also has a Post Graduate Diploma in Education (obtained from Makerere University in 2000). He has taught a range of subjects, including philosophy, ethics, human rights, and religious studies. He has published five books, 12 book chapters, and over 20 journal articles. Additionally, he has supervised several MA and PhD students to completion. His research interests are in African philosophy, ethics, political philosophy, and human rights.

Keba Cyril Muko obtained a Bachelor's degree in Philosophy and a Bachelor's degree in Sacred Theology from Demazod University in Kinshasa, DRC. He obtained a Master's degree in Philosophy at the Pontificia Universitas Gregoriana, and graduated with mention of Magna Cum Laude in 1999. Dr Muko lectured at St Joseph's Theological Institute, Cedara in South Africa. He obtained his PhD in Applied Ethics at the University of KwaZulu-Natal in 2017. His academic interests lie in the ethical analysis of the most cutting-edge issues relevant to human progress. Dr Muko is an Associate Member of the Mclean Center for Study of Culture & Values (Catholic University of America). He is currently a Lecturer at Saint Bonaventure's University College in Lusaka, Zambia.

Hupenyu Makusha holds a PhD in Ethics and is both a scholar and practitioner in the field of applied ethics, with a focus on migration, social justice, and African ethical perspectives. He teaches ethics at the School of Religion, Philosophy, and Classics at the University of KwaZulu-Natal. His academic interests encompass migration ethics, political ethics, and community development. In addition to his teaching and research, he coordinates refugee and migrant support services within the Catholic Archdiocese of Durban. Dr. Makusha has made a contribution to academic journals and conferences, advancing discussions on ethics, inclusion, and the ontology of diversity in Southern Africa.

Stephen Nkansah Morgan is a Senior Lecturer in the Department of Philosophy and Classics at the University of Ghana, where he teaches a diverse range of philosophy and ethics courses at both graduate and undergraduate levels. He holds a PhD in Ethics Studies from the University of KwaZulu-Natal in South Africa, as well as Master's and Bachelor's degrees in Philosophy, both from the University of Ghana. His research interests encompass African philosophy and applied ethics, focusing on environmental ethics, the ethics of technology, and artificial intelligence. He predominantly draws upon an African ontological and ethical understanding of nature, well-being, and harmonious living to critically reassess dominant concepts significantly influenced by Western and European worldviews.

Pascal Newbourne Mwale earned his PhD from the University of the Witwatersrand, Johannesburg. Dr Mwale is a Senior Lecturer in the Department of Philosophy, School of Humanities and Social Sciences, University of Malawi. He has a considerable track record of publications; some articles have appeared in internationally recognized peer-reviewed journals as well as a number of book chapters. His published articles in the *Nordic Journal of African Studies*, *Public Understanding of Science*, *Social Dynamics*, *Africa Development,* and the *Journal of Humanities,* as well as *Irinkerindo: A Journal of African Migration* (co-authored with Cobbener W. Sungani). Recent book chapters have appeared in the following volumes: *COVID-19 and Psychological Distress in Africa* (2024); *Ethics in Malawi* (2021); and *Babel Unbound* (2020). And there are also several book chapters in forthcoming volumes. His research interests are wide-ranging, with a focus on critical theory, media ethics, and phenomenology and existentialism, among others.

Nomatter Sande is a Research Fellow at the University of South Africa (UNISA). He is a cross-disciplinary researcher whose expertise lies at the intersection of religion, disability, migration, and pastoral ministry. The interdisciplinary nature of his research allowed him to synthesize insights from religious studies, theology, sociology, and anthropology, resulting in a holistic perspective on the role of religion in shaping societies undergoing profound changes.

Nothando Happy-Girl Shandu is a Lecturer in the Department of Ethics at the University of KwaZulu-Natal, where she is also pursuing her PhD. Her research interests include applied ethics, bioethics, environmental ethics, and African ethics, with a particular focus on the intersection between emerging biotechnologies and African ethical frameworks.

Cobbener W. Sungani was an MA student of African history at the University of Malawi. He held a Bachelor of Education degree with a major in History from the same institution. Cobbener was a Teaching Assistant in the Department of History, Archaeology, and Heritage Studies at the University of Malawi. His research interests were in the history of African media and colonial institutions. Some of his recent publications include, but are not limited to, works in peer-reviewed international journals, namely, *Irinkerindo: A Journal of African Migration* and *African Studies Quarterly.*In November 2024, Cobbener W. Sungani died after a short illness.

Lemohang Tebeli is a PhD Medicine-Bioethics candidate at the University of Cape Town, based in the Neuroscience Institute's Ethics Lab. His research interests span African philosophy, bioethics, transhumanism, and the ethics of artificial intelligence, with a particular focus on Afro-Communitarian conceptions of personhood and technological enhancement. He is also a Lecturer in the Department of Philosophy at Nelson Mandela University, appointed under the New Generation of Academics Programme (nGAP).

Part I
Foundations of African Population Ethics

Chapter 1
Introductory Chapter: Giving a Voice to African Ethics and Views in the Discourse on Population Ethics

Beatrice Okyere-Manu and Peter Mwipikeni

Abstract This chapter lays the foundation for this edited book, which critically explores population ethics through the lens of African ethics, history, and culture. It argues that Western ethical frameworks have dominated the discourse and debates on overpopulation and population policies. As a result, African ethical perspectives have been overlooked. The chapter explores concerns about overpopulation, from Malthus to the Club of Rome, arguing that these fears have often led to coercive population control measures. The chapter highlights the 1994 International Conference on Population and Development (Cairo), which emphasized reproductive rights and freedoms. This chapter sets the stage for the book's central contribution: to explore population issues from an African perspective, offering new insights for population policies rooted in African communal ethics, history, and cultural values.

Keywords Population ethics · Overpopulation · Poverty · High fertility

1.1 Introduction

Population ethics is concerned with the ethical issues that arise from the impact of human actions both on the people to be born and on the number of people who are born.[1] One of the human actions most feared by environmental activists in the West

[1] Arrhenius, G. Bykvist, K. Campbell, T., and Finneron-Burns, E. 2022. Introduction to The Oxford Handbook of Population Ethics. In The Oxford Handbook of Population Ethics, eds Arrhenius, Gustaf Krister Bykvist, Tim Campbell, and Elizabeth Finneron-Burns, 1–12. Oxford: Oxford University Press.

B. Okyere-Manu (✉) · P. Mwipikeni (✉)
Department of Philosophy and Ethics Studies, University of KwaZulu-Natal, Pietermaritzburg, KwaZulu-Natal, South Africa
e-mail: okyere-manv@ukzn.ac.za; MwipikeniP@ukzn.ac.za

B. Okyere-Manu, P. Mwipikeni (eds.), *An African Approach to Population Ethics*, Advancing Global Bioethics 23,
https://doi.org/10.1007/978-3-032-24464-2_1

is the high birth rates of poor women in the Global South.[2] These are castigated for causing overpopulation, environmental degradation, and poverty. This edited book comprehensively evaluates and analyzes ethically relevant issues of overpopulation and population policies from the perspective of African ethics, history, and culture. As suggested above, much of the analysis of population issues, such as overpopulation, migration, and population policies, has been undertaken from an evaluative framework shaped by Western ethical orientations. These frameworks include Ecocentric, libertarian, fairness, feminist, and consequentialist approaches. For example, welfare and happiness are fundamental concepts of Western utilitarian population ethics.[3] Similarly, libertarian and feminist population ethics emphasize procreative freedom, reproductive rights, and gender equality. However, to date indigenous sub-Saharan thought, which emphasizes community, harmonious relations, solidarity, identity, and the community's common good, has not been applied extensively to issues of population ethics.[4] In this edited book, we seek to contribute to the literature by examining and construing population issues that relate to Africans and the world from that very perspective of African ethics, culture, and history. Among other issues, we will examine how African ethical theories can be deployed to justify population policies. These policies may be used to interogate issues surrounding overpopulation.

The issue of overpopulation has influenced Western discourse on population policies and population ethics. This refers to the concept that there are too many people in a world of scarce resources. There is concern about the impact of excessive human population on the Earth's environment. There is also fear that high birth rates contribute to overpopulation and climate change.[5] This concern was perhaps first brought to the fore by the English social philosopher Thomas Malthus, who was concerned about the imbalance between a low rate of food production and a high population growth rate, which he was as causing poverty.[6]

In the twentieth century, Paul and Anne Ehrlich also expressed the fear that overpopulation could lead to an ecological catastrophe; and Garrett Hardin held that overpopulation could lead to the what he termed the "tragedy of the commons."[7] At a wider level, the Club of Rome maintained that overpopulation could lead to an

[2] Sasser, J. 2013. From Darkness into Light: Race, Population, and Environmental Advocacy. Antipode 00(0):1–18.

[3] Kolk, M. 2019. Demographic Theory and Population Ethics—Relationships between Population Size and Population Growth. The Institute for Futures Studies. Working Paper 11. 245

[4] See Metz 2022 on the main features of African communal ethics.

[5] Andersson, H. Brandstedt, E. and Torpman, O. 2021. Review article: The ethics of population policies. Critical Review of International Social and Political Philosophy 1–24.

[6] Andersson, H. Brandstedt, E. and Torpman, O. 2021. Review article: The ethics of population policies. Critical Review of International Social and Political Philosophy 1–24.

[7] Garret Hardin, The tragedy of the Commons. ***Science*** 13 Dec 1968 Vol 162, Issue 3859 pp. 1243–1248
https://doi.org/10.1126/science.162.3859.1243

ecological catastrophe.[8] Building on this, the concern and fears generated by discourses on overpopulation led to the development of remedies that included population control measures such as coercive population policies. For instance, Ehrlich proposed adding doses of sterilant to water supplies and staple foods.[9] In another contribution, Onora O'Neill highlighted how coercive and non-coercive population policies can be used to mitigate the threat of famine, which is exacerbated by the prevalent high rates of population growth.[10] Coercive population policies include sterilization; non-coercive population policies include public campaigns that promote the use of contraceptives. This discourse, linking overpopulation with resource scarcity and ecological catastrophe, is an aspect of the important background that has shaped the emergence of population ethics.

It is also the case that the discourse of overpopulation emerged in a racialized world, one which is characterized by global coloniality. A key structuring principle of the world is the coloniality of "being," which denied the humanity of Africans.[11] Africans and non-Europeans were othered as ontologically flawed beings who were lacking in rationality. This led to their invention as beings constituted by deficits in terms of their writing, history, development, and democracy.[12] Additionally, global coloniality bifurcated the world into the Global North zone of being and the Global South zone of non-being. The problem of overpopulation was thus linked to poor people from the Global South, who irrationally proliferated exponentially in an unsustainable way. Hardin's fears of overpopulation were shaped by race.[13] For him, it was the people in the Global South who contributed to overpopulation, and he advocated population management measures to curb reproduction.[14]

The overpopulation discourse, with its orientation toward recommending coercive population control policies that undermined individual rights, sparked outrage. The concern, as suggested above, was that the overpopulation discourse racialized the poor people in the Global South as irrational beings who contributed to overpopulation and resource scarcity. This led to the staging of the International Conference on Population and Development in Cairo in 1994, which attempt to

---

[8] Andersson, H. Brandstedt, E. and Torpman, O. 2021. Review article: The ethics of population policies. Critical Review of International Social and Political Philosophy 1–24.

[9] Andersson, H. Brandstedt, E. and Torpman, O. 2021. Review article: The ethics of population policies. Critical Review of International Social and Political Philosophy 1–24.

[10] Andersson, H. Brandstedt, E. and Torpman, O. 2021. Review article: The ethics of population policies. Critical Review of International Social and Political Philosophy 1–24.

[11] Maldonado-Torres, N. 2007. On the coloniality of being. Cultural Studies 2–3: 240–270.

[12] Grosfoguel, R. 2011. Decolonizing Post-Colonial Studies and Paradigms of Political Economy: Transmodernity, Decolonial Thinking, and Global Coloniality. Transmodernity: Journal of Peripheral Cultural Production of the Luso-Hispanic World 1:1 38. 7

[13] Andersson, H. Brandstedt, E. and Torpman, O. 2021. Review article: The ethics of population policies. Critical Review of International Social and Political Philosophy 1–24.

[14] Beatrice Okyere-Manu (2016) Overpopulation and the Lifeboat Metaphor: A Critique from an African Worldview, International Studies in the Philosophy of Science, 30:3, 279–289, DOI: 10.1080/02698595.2017.1316114. 280

eschew coercive population policies. Principle 8 in the Program of Action stipulates that:

> Everyone has the right to the enjoyment of the highest attainable standard of physical and mental health. States should take all appropriate measures to ensure, on a basis of equality of men and women, universal access to healthcare services, including those related to reproductive health care, which includes family planning and sexual health. Reproductive health-care programmes should provide the widest range of services without any form of coercion. All couples and individuals have the basic right to decide freely and responsibly the number and spacing of their children and to have the information, education, and means to do so.[15]

This principle emphasizes couples' right to decide freely and responsibly the number and spacing of their children. From the above discussion, it is clear that three critical features shaped the emergence of population ethics in the West: overpopulation, race, and the promotion of reproductive rights. The ecocentric theory emphasizes actions that are morally right and seek to prevent overpopulation, which is blamed for contributing to an ecological disaster (Andersson et al. 2021). The consequentialist theory addresses the issue of overpopulation and seeks to prevent a life of misery.[16] A miserable existence can be caused by resource scarcity resulting from overpopulation. The fairness approach is concerned that procreation brings into existence children who create a public burden, as they can contribute to the worsening of climate change. The concern is whether the costs and burden of procreation should be shared with others, or whether they should be solely the responsibility of parents. The libertarian approach to population issues is shaped by the quest to promote reproductive rights.[17] The feminist ethical approach is applied to population issues to promote procreative freedom and negative and positive reproductive rights. Additionally, the feminist approach to reproductive rights represents an epistemological rupture with the traditional epistemology of overpopulation and resource scarcity. Such an epistemology blames overpopulation in the Global South for causing poverty (Schultz 2018). Moreover, the racialized epistemology contributes to the "othering" of poor young women in the global South as irrational, dangerous bodies, causing overpopulation, poverty, and ecological catastrophe. According to feminist ethics, the global capitalist system contributes to poverty and environmental degradation.[18]

The lacuna in the discourses that use the above-mentioned ethical theories to analyze and evaluate issues of overpopulation and population policies is the silence around African ethics, culture, and values. These discourses are all grounded in

[15] Andersson, H. Brandstedt, E. and Torpman, O. 2021. Review article: The ethics of population policies. Critical Review of International Social and Political Philosophy 1–24. 5.

[16] Andersson, H. Brandstedt, E. and Torpman, O. 2021. Review article: The ethics of population policies. Critical Review of International Social and Political Philosophy 1–24.

[17] Andersson, H. Brandstedt, E. and Torpman, O. 2021. Review article: The ethics of population policies. Critical Review of International Social and Political Philosophy 1–24.

[18] Andersson, H. Brandstedt, E. and Torpman, O. 2021. Review article: The ethics of population policies. Critical Review of International Social and Political Philosophy 1–24.

Western moral theories. There is, at present, a deficiency of work that analyses population issues through the lenses of Indigenous African communal ethics. Much theoretical reflection on overpopulation and population policies is heavily informed by a Western individualistic theoretical orientation that pervades most such ethical orientations, such as libertarian, consequentialist, and fairness approaches. In contrast, the Western feminist theoretical framework does have a community orientation. However, it does not encompass all the communitarian features of the African theoretical framework. There are few scholarly works that have engaged with the application of African ethics and culture in the discourse of evaluating issues associated with overpopulation and population policies.

Additionally, there has been little critical engagement with African perspective on the population issues in the light of African history from the time of the slave trade to the era of colonialism, neocolonialism, and neoliberalism. The chapters in this edited book are significant as they are part of an ongoing project to develop population ethics from the perspectives of the indigenous ethical orientations, history, and cultural values of Africans, in a manner that responds specifically to African needs, interests, and priorities.

1.2 The Need for an African Perspective on Population Ethics

This project is important for several reasons. One of these is that the African voice is underrepresented in the field of population ethics, which does not contribute to the field's enrichment. This leads to the uncovering of distortions and implausible assumptions that result from the imposition of a Eurocentric perspective on pertinent global demographic issues. There is a possibility of error and faulty assumptions if population ethics are grounded solely in the Western ethical tradition. Western ethics focuses on promoting individual interests and prioritizing the interests of people from the West. Such a population ethics can misrepresent and undermine the interests of Africans and other non-Western people. The African voice in population ethics is crucial for promoting the interests of Africans and enriching the field. In one contribution, Behrens (2017) cited Godfrey Tangwa, a Cameroonian philosopher, who observes that "reality and human experience are too vast and too diverse to be captured by one culture or conceptual paradigm, be it globally dominant or not."[19] This book contributes to a more comprehensive understanding of population ethics by adding the African voice. It thereby marks the end of the marginalization of African knowledge in the field of population ethics and the enrichment of the field. It offers a novel approach to evaluating and analyzing population issues from the perspective of African ethics, culture, and history, intending to

[19] Behrens, K.G. 2017. Hearing sub-Saharan African voices in bioethics. Theor Med Bioeth 38:95 99. 97

contribute new knowledge to population discourses by drawing on these continent-wide perspectives to examine issues such as overpopulation and population policies. Among the questions that will be tackled are the following:

Can population policies be justified from the perspective of African ethical orientation?

Is it plausible to propose the notion of an "African" population ethics?

How can we consider the notion of overpopulation in the light of African history, culture, and ethics?

This book is also important in terms of contributing to the attainment of epistemic freedom. Issues of population, such as overpopulation and population policies, are influenced by colonial epistemologies that promote Western interests. There are colonial demographic epistemologies that reduce the problem of poverty to one of overpopulation in the Global South.[20] Such a framing promotes Western interests by shifting the blame from colonialism and global practices of capitalist accumulation. Colonial epistemologies regard African knowledge as "inferior." This book contributes to epistemic freedom by attempting to analyze and evaluate the continent's population issues from an African perspective. Epistemic freedom involves an attempt to think and theorize about population issues in Africa. Benyera (2022), cited the words of Sabelo Ndlovu-Gatsheni, who holds that "epistemic freedom speaks to cognitive justice. Epistemic freedom is fundamentally about the right to think, theorize, interpret the world, develop one's own methodologies, and write from where one is located and unencumbered by Eurocentrism."[21] This book regards Africa as the locus of enunciation for population issues that pertain to the continent.

1.3 The Shape and Contents of the Book

The book is divided into four parts. Before the main parts, the introductory chapter, "Introduction: Giving a Voice to African Ethics and Views in the Discourse on Population Ethics," by Beatrice Okyere-Manu, Peter Mwipikeni, lays the groundwork for the rest of the volume. It introduces the core concepts, debates, and ethical frameworks that inform African perspectives on population issues and sets the stage by emphasizing the importance of African voices and challenging Western-centric approaches..

Part I of the book focuses on Population Ethics from an African Perspective. It contains two chapters. Chap. 2, "African Ethics and Population Issues" by Wilfred Lajul, questions the neo-Malthusian idea that population growth causes poverty, environmental degradation, and climate change. Such an idea is problematic,

[20] Schultz, S. 2018. Demographic futurity: How statistical assumption politics shape immigration policy rationales in Germany, Susanne Schultz. Society and Space 3 7(4): 644–662.

[21] Benyera, E. 2022. On the Elusiveness of Epistemic Freedom in Africa and the Need to Unmask the Colonial Contract. In Africa and the Fourth Industrial Revolution: Curse or Cure? Ed. Everisto Benyera, 1–15. Cham: Springer.

considering that developing countries have higher rates of population growth, while developed countries have lower rates. It implies that overpopulation, environmental degradation, and climate change are caused by problematic high fertility rates and birth rates in the Global South. Lajul shows that it is, in fact, overdevelopment and the impact of modern technological advancements that depletes resources, damages the environment, and contributes to climate change. Lajul deploys an African ethical perspective to generate a more nuanced view of the relationship between population and human behavior at a global level.

Chap. 3 by Wilfred Lajul, titled "Meaning of Life and Population Issues from an African Perspective," involves using the African notion of the meaning of life to construe African Population Ethics that value the number of human people born without consideration of the circumstances of birth and how they will live after they are born. Lajul contrasts this African perspective with the Western view which promotes the human duty to determine who is born and how many people are to be born. Lajul's chapter questions whether it is plausible to construe the meaning of life as dependent on the Western perspective or the African perspective, and he uses a critical analytical method to explore the implications of these divergent views in population discourse. The author questions the notion that the meaning of life can be considered in relation to who is born and how many people are born in this world.

The second part of the book explores the rich African cultural and philosophical traditions that shape views the continent's on procreation, family, and community. It examines how these traditions influence the ethical considerations related to population dynamics, providing a deeper understanding of the values and beliefs that underpin African approaches. The following chapters are to be found in this part of the book.:

Beatrice Okyere-Manu's chapter discusses the African proverbs from the Akan people in Ghana which encourage high birth rates, as well as those that advocate for small family sizes when resources are inadequate. Okyere-Manu demonstrates that proverbs promoting high fertility rates are popular and cherished among Africans. She questions the desirability of these proverbs, which promote high fertility rates of women when there are high levels of poverty and environmental degradation. She uses notions of Western consequentialism to critique the desirability of such proverbs.

Next Sande Nomatter's chapter considers "Tension between Procreation as a Duty and Individual Right in African Population Ethics." Drawing on ethnographic examples from various Shona, Yoruba, Dagomba, Maasai, and Akan teachings, the chapter illustrates how procreation serves not only to sustain family and lineage but also to uphold social order and ancestral continuity. It contrasts these indigenous perspectives with emerging reproductive rights discourses across the continent, highlighting the growing tensions between tradition and modernity. Through the conceptual lens of moral demography, the study reveals that demographic behavior in Africa cannot be disentangled from local moral economies, historical influences such as colonization and Christianity, and the evolving roles of institutions like the church and feminist movements.

The final chapter in this part is by Makusha Hupenyu. This contribution examines the ethical dimensions of procreation in the context of climate change, with a particular emphasis on the African perspective. It questions the morality of having children amidst escalating environmental challenges, juxtaposing individual reproductive rights with collective ecological responsibilities. The chapter highlights Africa's vulnerability to climate impacts and addresses issues of intergenerational justice, resource sustainability, and the potential hardships faced by future generations. Hupenyu examines ethical frameworks for sustainable reproductive choices, assessing population control, climate education, and counterarguments that favour procreative freedom and technological solutions. It seeks a balanced, inclusive dialogue, advocating for decisions prioritizing the planet's long-term wellbeing while respecting Africa's unique socio-cultural and economic contexts.

Part III examines the real-world challenges and implications of population growth, distribution, and change in Africa. It addresses critical issues such as environmental sustainability, resource management, migration, and the ethical dilemmas associated with population control measures.

The first chapter in this part, by Peter Mwipikeni, chapter shows that the overpopulation crisis is a historically contingent effect of global capitalist accumulation practices. For Mwipikeni, the overpopulation crisis in Africa, associated with a Malthusian negative relationship between population growth and welfare, is a result of global capitalist accumulation practices that have drained value from Africa throughout the colonial and the postcolonial eras. Mwipikeni demonstrates how ethical ideas associated with the Sotho aphorism, "*feta kgomo o tshware motho*," ("go past the cow and catch the human being" is an ethical maxim in the African philosophy of Ubuntu among the Bantu-speaking peoples.) (Ramose 2010) can reconfigure a world economy that would address the overpopulation crisis in Africa. This is an aphorism that prioritizes the promotion of the welfare and life of all human beings above the accumulation of wealth.

The next contribution, by Stephen Nkansah Morgan, examines the ethical need to manage human population growth for environmental sustainability, using African communitarian ethics as a guiding framework. It contrasts this approach with Western theories—such as utilitarianism, deontology, and rights-based ethics—highlighting their insights and shortcomings, particularly in relation to intergenerational justice and environmental concerns. The chapter examines the advantages and disadvantages of population control, striking a balance between ecological concerns and the ethical risks of coercion. It concludes that African communitarianism, with its emphasis on interconnectedness, shared responsibility, cultural values, and concern for future generations, offers a more holistic and ethically sensitive approach to addressing population and environmental challenges.

Keba Muko's chapter aims to demonstrate the significance of reflexive self-consciousness and African intersubjectivity in reestablishing the existential connection to the environment, which is crucial for mitigating the adverse effects of global warming on the wellbeing of African populations. Muko employs African ethical analysis and objective argumentation to contribute to the understanding of population ethics issues influenced by African ethics of intersubjectivity, which regard the environment as a shared good for the wellbeing of African populations, present and future.

Next, Mwipikeni shows the limitations of scholarly work that discusses the reconfiguration of population control after the 1994 Cairo Conference, which supposedly eschewed population control policies. He discusses how Bhatia et al. identify three forms of reconfiguration of population control: demopopulationism, biopopulationism, and geopopulationism and considers that Amanda Shaw and Kalpana Wilson's argument that population control can take the form of necropopulationism. In response, Mwipikeni shows another form of population control in Africa, which he calls necroeconomic-populationism.

Part IV provides concrete examples of population ethics in specific African contexts. It presents case studies that explore contemporary issues, such as youth perceptions of population growth, resource scarcity, xenophobia, and migration, offering valuable insights into the complexities and nuances of population dynamics on the continent.

Shandu and Okyere-Manu's chapter explores young people's perceptions of human population growth in contemporary South Africa, using a population sample of university students from the University of KwaZulu-Natal. The authors explore the views of young South Africans on a range of issues: whether there is actually overpopulation; their preferences regarding family size; their perspectives on population growth and environmental degradation; the effects of population growth on future generations; and socio-cultural norms that promote high fertility rates, family planning, and population growth management. The chapter shows that these students have different views on these issues of population studies. The authors also discuss the Child Grant, a South African government social welfare policy. According to the authors, the Child Grant could be regarded as an incentive for poor young girls to have children, thereby again contributing to population growth.

The next contribution, Pascal Newbourne Mwale and Cobbener W. Sungani's "Population Boom versus Scarcity of Resources in Malawi: An African Historical-Ethical Critique," shows that there is both high population growth rate and unequal distribution of wealth in Malawi. They demonstrate how such conditions can lead to poverty and undermine the welfare of a significant portion of the population. Mwale and Sungani advocate a substantial redistribution of wealth to promote the welfare of the poor. They use the African concept of Ubuntu to argue that wealthy individuals, the government, and international humanitarian organizations should promote the welfare of the poor population in this country.

Finally, the chapter by Lemohang Tebeli views the issue of migration and xenophobia in population discourse through the lens of African communitarian ethics. It holds that the African foreign migrants are regarded as causing overpopulation in South Africa, and competition both in the job market, and with regard to scarce public resources. He questions the overpopulation thesis in migration discourse and links socio-economic problems to colonialism. Tebeli suggests that the problems of xenophobia can be addressed by African communal ethics, which promote solidarity and the inclusion of foreigners within a collective identity.

References

Andersson, H., E. Brandstedt, and O. Torpman. 2021. Review article: The ethics of population policies. *Critical Review of International Social and Political Philosophy*1–24.

Arrhenius, G., K. Bykvist, T. Campbell, and E. Finneron-Burns. 2022. Introduction to the oxford handbook of population ethics. In *The oxford handbook of population ethics*, ed. Gustaf Krister Arrhenius, Tim Bykvist, Campbell, and Elizabeth Finneron-Burns, 1–12. Oxford: Oxford University Press.

Behrens, K. G. 2017. Hearing sub-Saharan African voices in bioethics. *Theor Med Bioeth* 38:95–99.

Benyera, E. 2022. On the elusiveness of epistemic freedom in Africa and the need to unmask the colonial contract. In *Africa and the Fourth Industrial Revolution: Curse or Cure?* ed. Everisto Benyera, 1–15. Cham: Springer.

Grosfoguel, R. 2011. Decolonizing post-colonial studies and paradigms of political economy: Transmodernity, decolonial thinking, and global coloniality. *Transmodernity: Journal of Peripheral Cultural Production of the Luso-Hispanic World* 1 (1): 38.

Kolk, M. 2019. Demographic theory and population ethics–relationships between population size and population growth. *The Institute for Futures Studies*. Working Paper 11.

Maldonado-Torres, N. 2007. On the coloniality of being. *Cultural Studies* 2–3:240–270.

Metz, Thadeus. 2022. *A relational moral theory: African ethics in and beyond the continent*. Oxford: Oxford University Press.

Okyere-Manu, Beatrice. 2016. Overpopulation and the lifeboat metaphor: A critique from an African worldview. *International Studies in the Philosophy of Science* 30 (3): 279–289. https://doi.org/10.1080/02698595.2017.1316114.

Ramose, Mogobe B. 2010. The Death of Democracy and the Resurrection of Timocracy. *Journal of Moral Education* 39 (3): 291–303.

Sasser, J. 2013. From darkness into light: Race, population, and environmental advocacy. *Antipode* 00 (0): 1–18.

Schultz, S. 2018. Demographic futurity: How statistical assumption politics shape immigration policy rationales in Germany, Susanne Schultz. *Society and Space 3* 7(4): 644–662.

Chapter 2
African Ethics and Population Issues

Wilfred Lajul

Abstract Population issues are about public discussions, debates, and depictions surrounding population issues, including population growth, fertility rates, and their social, economic, and environmental implications. This discourse is centered on the ways in which populations, especially their size, growth, and composition, are interpreted and understood within society. Such interpretations often influence policies and actions related to development, health, and social issues. African ethics, on the other hand, is a field of reflective exploration of the moral beliefs, values, and practices of people indigenous to sub-Saharan Africa, emphasizing community, interconnectedness, and the common good. On the basis of these values, an African would have a different understanding and interpretation of the issues surrounding population composition, growth, and distribution of resources among humans. Malthus posits that population growth outstrips resource production, leading to inevitable poverty, famine, and conflict, unless checked by "preventive" or "positive" measures. The implication is that poverty, famine, and conflicts are caused by an increase in population. But the increase in population is not uniform across the world. This is because developed countries have lower rates of population growth, while developing countries have higher rates. This implies that the problem of overpopulation is coming from the developing countries. This same argument can be repeated for both famine and conflict. Such arguments completely ignore the damaging effects of overdevelopment and the impact of modern technological advancements, which cause global warming, depleting limited natural resources, and exacerbating the level of suffering for humanity, especially in developing countries. Using a critical analytical method, this chapter investigates the African ethical perspective of this discourse with the hope of generating a more balanced view on the relationship between population and human behavior in the world.

Keywords Africa · Population discourse · Ethics · Development

---

W. Lajul (✉)
Gulu University, Religious Studies and Philosophy, Faculty of Education and Humanities, Gulu, Uganda

B. Okyere-Manu, P. Mwipikeni (eds.), *An African Approach to Population Ethics*, Advancing Global Bioethics 23,
https://doi.org/10.1007/978-3-032-24464-2_2

2.1 Introduction

Since its inception, there has been much debate around the notion of population discourse. Its initial discussion by Malthus (1798) was based on the principle that population growth always outstrips resource production, leading to inevitable poverty, famine, and conflict, unless this is checked by preventive or positive measures. This was later compounded by the 1968 publication of Paul Ehrlich's book, *The Population Bomb.* After this, as Coole (2021) contends, the implications of this discourse reached toxic levels, generating negativity about the discourse and causing many analysts to abandon the discussion. Coole underlines that "During the 1960s and 70s, reducing human numbers was embraced as integral to radical social transformation. Subsequently, such neo-Malthusian prescriptions became so toxic that they disappeared from the political agenda. Only recently has the issue resurfaced."[1]

This chapter discusses the different perceptions and theories underlying population control. The main reason for population control, it is argued, is to balance human development with human population sizes. Our problem with this approach is that it ignores other equally important measures to ensure balance in development in terms of a better quality of human life, in view of the available resources in the world. We are well aware that the human population is generally increasing, and that the size of our universe remains the same. The conclusion leading from this statement is that the world's resources are insufficient for the citizens of the world because of the increase in population alone. This conclusion, and its implications, ignore the fact that the insufficiency of both our resources and the space within which humans can live comfortably is affected not only by growth in population, but also by industrial pollution and the kind of technology being used to exploit the natural resources. Realistically, environmental pollution and damage impacting the available spaces for human existence are caused by overdevelopment and the bad use of modern technologies. Therefore, shifting blame to population increase alone does not help to solve the problem.

More importantly, population control is targeting poor countries, whose rate of fertility is up to 4.7 per woman, while the developed countries, which have the greatest responsibility for polluting the human environment due to their higher levels of technology and intensive development programs, have a fertility rate of 1.7. Apparently, all of the measures currently put in place for population control are targeting the poor countries. This is what is most disturbing, and requires that such discussions be revisited.

This chapter, then, investigates the African perspective of this discourse, aiming to draw attention to other equally important issues besides population control. We shall do this by first discussing the general issues surrounding population ethics, and

[1] Diana Coole, "The Toxification of Population Discourse: A Genealogical Study," *The Journal of Development Studies* 57, no. 9 (2021): 1454–1469, https://doi.org/10.1080/00220388.2021.191547

then Western population ethics, before analysing the African population ethical perspectives.

2.2 Population Ethics

Population ethics is an area of applied ethics concerned with the ethical problems caused by human activities aimed at controlling who is born and how many people are born in the future.[2] In the introduction to the *Oxford Handbook of Population Ethics*, population ethics is "the investigation of various ethical issues that arise when we consider how our actions affect both who is born and how many people are born."[3]

Population ethics can be regarded as the study of human activities in as far as they affect who is born and how many people are born. This study is usually divided into two forms: the axiological and the normative.[4] The difference between the two is that the axiological form considers "[…] how we ought to value different populations, whereas the normative form considers how we ought to act in relation to populations, for instance, which populations we ought to bring about."[5] This means that, from the axiological point of view, population ethics studies the value judgments we make with regard to different human populations, while from the normative perspective, we consider the norms and policies that regulate future human populations.

The central problem with both of these approaches lies in human attitude and behavior in terms of value judgments and regulatory norms in respect of those yet to be born. The basis of such value judgments and norms, from the ethical point of view, is always problematic. To illustrate such value judgments from the axiological point of view, Gustaf et al. (2022) maintain that population axiology attempts to determine how to rank human populations.[6]

For instance, would it be better if one excellent life existed than if two different decent lives existed? Do we make the world better by creating more happy people? Is a world in which many people live a very low-quality life better or worse than a world in which a very small number live a very high-quality life?[7]

The answers to these questions are those that create a big moral problem. This is because, centrally, it questions whether or not the quality of life or human well-being is the essence of human happiness, or whether there is something greater than

---

[2] Henk ten Have and Maria do Céu Patrão Neves, "Population Ethics," in *Dictionary of Global Bioethics*, (Cham: Springer, 2021), https://doi.org/10.1007/978-3-030-54161-3_410

[3] Gustaf Arrhenius, Krister Bykvist, Tim Campbell, and Elizabeth Finneron-Burns, eds., *The Oxford Handbook of Population Ethics* (Oxford: Oxford University Press, 2022)

[4] Gustaf, *The Oxford Handbook of Population Ethics*.

[5] Gustaf, *The Oxford Handbook of Population Ethics*, 2.

[6] Gustaf, *The Oxford Handbook of Population Ethics*, 2.

[7] Gustaf, *The Oxford Handbook of Population Ethics*, 2.

human well-being. If human well-being is to be considered the essence of the meaning of human life, then life is a commodity that is at the whims of humans to manipulate and shape in whatever way they want, since it can be shaped by human activities. If the contrary is true, however, then there is something more than mere human well-being or quality of life that creates human happiness and the meaning of life that comes with it.

This is exactly the concern of this chapter: to find what constitutes meaningful human life as a basis for making value judgments about human populations. This also means it might not be quite obvious that, because there is a high quality of life or human well-being among the members of a given human population, then there is greater meaning in the life of such a human population. The other factors that might equally be important besides human well-being may need to be investigated to determine the proper basis of value judgment about human populations, which is population ethics.

The normative population ethical perspective asks questions like: Do we do something wrong when we create a person with a decent life when we could instead have created a different person with an excellent life? Do we have an obligation to create more happy people? Do we have a duty to create many people with a very low quality of life rather than create a few people with a very high quality of life?

In answer to these questions, the views are again divided. Western ethics, with its traditional ethical theories such as utilitarianism, Kantianism, and contractualism, sees it is a human duty to regulate human populations and the quality of life within them. This is done through the utility maximization principle (utilitarianism), or on notions such as duty for duty's sake (Kantianism). Utilitarianism would propose policies and regulatory measures that would reduce the population of low-quality lives, in order to maximize the well-being of populations with high-quality lives. And it is a human duty to do that, not for any particular benefit, but because it is right in itself (Kantianism). Besides, we do not have any contract with the future generations to take care of their interests, because such a contract would not benefit the present generations, since the need to improve the quality of life of the present generation lies squarely with them.

African metaphysics, by contrast, would insist that we are duty-bound, not only to fellow humans, past, present, and those that will come in the future, but also to nature, which is our environment, the governing laws of nature created and governed by God. In this philosophical view, humans cannot make laws or policies that would injure others (present and future), the environment (nature), and human relationships with God.

In the final analysis, we can deduce that simple forms of consequentialism would claim that we ought to bring about the best population without considering other, more complex theories about axiological and normative aspects of population ethics. The effect of our actions on the present generation has direct consequences on the well-being and quality of life today, and not necessarily tomorrow. However, answers to these questions matter because they affect the moral standing of future people in the decisions we currently make.

Some proponents of the Western philosophy outlined here think that "Future people pose an especially hard problem for our decision-making, since their number and identities are not fixed; rather, they depend on the choices we make."[8] The ethical question is: what is the basis of thinking in this way? Just because the number and identities of the future people are not fixed, does not mean we can take them for granted. The future people matter too for the sustenance of future humanity. The wealth and affluence of the present generation would be equally meaningless if there is no guarantee that they are sustainable. If they are to be sustainable, then they would need future people. The future people also have the right to exist.

2.3 Western Population Ethics

One of the Western ethical theories that is usually employed to discuss population ethics is known as "welfarism." This theory is derived from Jeremy Bentham's statement that "The greatest happiness of the greatest number is the foundation of morals and legislation."[9] While the term welfarism was originally coined by the British economist John Hicks,[10] it has more recently been introduced into contemporary debates about social justice by Amartya Sen.[11]

Welfarism theory demands that "we judge the state of a society in general, and just distributions in particular, by the welfare of the people concerned, and by their welfare only."[12] But welfare is an emotional state, subjectively experienced by an individual. It is more or less synonymous with the terms well-being, happiness, desire fulfilment, and utility. Utility (from Latin "utilis": useful) is the preferred term in economic analysis, and it is also the root of the word "utilitarianism.".

In connection with population ethics, it is argued that the greatest happiness of the greatest number is determined by the quality of life and access to welfare. It is further proposed that human welfare is undermined by an uncontrolled population explosion. This was the key argument of Paul Ehrlich in his 1968 book *The Population Bomb*, in which he insinuated that population increase is dialectically counterproductive to human well-being or welfarism.[13]

Criticizing this understanding, University of Oslo (UIO) observes that "It is far from obvious that my welfare can only be increased at the expense of the welfare of

[8] Gustaf, *The Oxford Handbook of Population Ethics*, 2.

[9] Jeremy, Bentham. *An Introduction to the Principles of Morals and Legislation*, edited by J. H. Burns and H. L. A. Hart (The Athlone Press. 1970).

[10] John R. Hicks, *Wealth and Welfare: Vol. I of Collected Essays in Economic Theory* (Oxford: Basil Blackwell, 1981), 136.

[11] Amartya Sen, *The Idea of Justice* (Cambridge, MA: Belknap Press of Harvard University Press, 2009)

[12] University of Oslo -ui)

[13] R. Paul Ehrlich, *The Population Bomb*, 12th ed. (New York: Ballantine Books, 1968).

another person."[14] On the other hand, it is also obvious that welfare is influenced by economic goods. Man does not live by bread alone, but the welfare of a hungry person increases when she is given bread to eat or money to buy bread. Thus, the distribution of economic goods is, to some extent, a distribution of welfare. Welfarism implies that the distribution of economic goods is evaluated by the effect it has on the welfare of the persons concerned.

Picking up on this point, Slesnick (1998) argues that the measurement of welfare forms the foundation of public policy analysis. "A full consideration of taxes, subsidies, transfer programs, health care reform, regulation, environmental policy, the social security system, and educational reform must ultimately address the question of how these policies affect the well-being of individuals."[15]

Ordinarily, the above policy interest looks healthy, but the question arises when considerations of welfare are linked to the question of population increase and the quality of human life. The thinking here is that the reason why welfare is scarce is because of population increase, and that thus the solution to the scarcity of welfare lies in the systematic reduction of population. The idea here is that it is better to have few in the world with high levels of well-being than to have many human beings with low standards of human well-being. Given that well-being and human happiness are synonymous, it is argued, the higher the quality of life, the happier a given human population.

It is here that the problem lies, in linking human welfare and human happiness; consequently, preferring higher qualities of life to lower ones, and making policies that systematically reduce poor populations in favour of affluent ones.

In 2019, Chomanski published an article, "What's wrong with designing people to serve?," in which it was stated that there is nothing wrong with designing people who can serve as servants. But the idea of designing people who can serve as servants or workers comes directly from Western population ethics. The conviction behind this ethics is that to maintain a high level of human well-being and quality of life, we do not need more human beings, but rather Artificial Persons (APs), who can serve as servants or workers. This is because to sustain a high standard of human welfare, humans constantly need servants or workers.

In response to this practical proposal, Musiał (2022) asked, "Can we design artificial persons without being manipulative?" The answer to this question seems to be no. When Musiał asked, "Does designing such AP workers involve a manipulativeness on the side of designers?," he answered his own question by saying, "I believe it does."[16]

---

[14] Oslo. UIO, https://www.uio.no/studier/emner/sv/oekonomi/ECON1220/h04/undervisningsmateriale/deadweight.pdf

[15] Slesnick, T. Daniel, (1998, 2108). "Empirical Approaches to the Measurement of Welfare," *Journal of Economic Literature, American Economic Association*, vol. 36(4), pages 2108–2165. 2108. Slesnick 1998, 2108

[16] Maciej Musiał, "Can We Design Artificial Persons without Being Manipulative?" *AI & Society* (2022). https://doi.org/10.1007/s00146-022-01575-z

Musiał continues to observe that manipulativeness means "willingness to orchestrate another's choices in an excessive manner" by "limiting another's options" and "disregarding their rational capacities." I contend that, in the case of AP workers, designers attempt to lead APs to act as they want them to."[17]

Musiał concludes that manipulativeness is present in the design of any Artificial Person (AP) or Artificial Intelligence (AI), even if the designed desires are not overwhelming and are accompanied by other non-designed desires.[18] This seems to be compatible with Chomanski's assertion that manipulativeness is present even if designers "do not exercise complete control over the AI's psychological profile."[19]

Our argument for this is that behind the apparent purely innovative scientific reasons for designing AP or AI, the hidden motive is the reduction of natural human beings who have the propensity to overpopulate, thereby undermining the quality of life in this world. This is the practical application of utilitarian ethics, that seeks the maximization of the highest quality of life for the greatest number. In this case, however, the majority they are referring to is not the majority of humanity, but the majority of those whose high quality of life that can easily be compromised by an increase in poverty-stricken populations. Their argument revolves around the fact that we do not want human beings who can produce children and grandchildren, but rather robots that have no capacity to reproduce other robots on their own.

In the view of this chapter, the biggest problem with Western population ethics is the comparison of human life with the existence of artificial persons. The fact that an AP or AI can act, work, think (if that is correct), serve or work like a human being does not guarantee that they have life in the sense contained in human beings. From the point of view of the performance of tasks that humans can do, AP or AI can even be more proficient in terms of efficiency and accuracy. But they cannot have life in the sense that we understand human beings to have it.

Josef de Varies defines life as "immanent activity (*action immanens*) in contrast to the kind of activity that is directed to the transformation of other things (*action transiens*)."[20] Life, according to de Varies, is an inner experience in conscious seeing, thinking, feeling, and striving, depending on the relative intensity of these activities on the freshness and power of the constituting organs. Josef de Varies adds that, "life reveals itself as a constant becoming and an inexhaustible, many-faceted self-development that proceeds from the inside; this stands in sharp contrast to the rigidity and opaqueness of lifeless bodies, especially machines."[21] Centrally, de Varies concludes that life is "a mysterious, creative power that has been coursing through the world since primeval times; … life as an essentially irrational power

---

[17] Ibid.

[18] Ibid.

[19] Bartosz Chomanski, "What's Wrong with Designing People to Serve?" *Ethical Theory and Moral Practice* 22 (2019): 993–1015, https://doi.org/10.1007/s10677-019-10029-3

[20] Josef de Vries, "Life." In *Dictionary of Philosophy*, edited by Walter Brugger. Translated and edited by Kenneth Baker. Spokane, WA: Gonzaga University Press, 1972, 223.

[21] Ibid.

that escapes all conceptual definition; [... is] constant becoming which is opposed to immutability of existence."[22]

There is a big question as to whether AP or AI can have life in the sense described above. In addition, thinking, in our view, is the ability of conscious self-reflection; and we may think that APs and AIs may be designed to have some degrees of self-reflection, but whether such activity is a conscious one in the sense proper to humans has not been conclusively established. According to Lewis and Sarkadi (2023), this is impossible. They write:

> Artificial Intelligence (AI) is about making computers that do the sorts of things that minds can do [...]. However, AI systems usually do these tasks with an unusual imbalance of insight and understanding: new, deeper insights are present, yet many important qualities that a human mind would have previously brought to the activity are utterly absent [...]. One core feature that humans bring to tasks, when dealing with the ambiguity, emergent knowledge, and social context presented by the world, is reflection. Yet this capability is utterly missing from current mainstream AI.[23]

Others who doubt that possibility are authors like Hildt, who maintain that "We need to talk less about artificial consciousness and we need to talk more about the lack of consciousness in current robots and AI."[24] On the one hand, many prominent scholars argue that AIs, like machinery in general, have no moral status.[25] On the other hand, there are also people who argue that, if an AI has perceptual characteristics, it must have moral status.[26]

These divergences are not the reasons we bring APs and AIs into the discourse on population ethics. What we are saying here is that utilitarian Western population ethicists think the use of APs and AIs is an innocent way of regulating the human population, without hurting anybody. For them, it is a surer way of guaranteeing the services and workers humans need to keep high standards of life without permitting an increase in welfare competitors in terms of additional human beings. Moreover, these additional human beings have no guarantee that they will have decent welfare to live by. What they are not saying is that, by introducing this as a policy, they are controlling and determining who is born and how many people are born.

---

[22] Ibid.

[23] Peter R. Lewis and Stefan Sarkadi, *Reflective Artificial Intelligence* (2023), 1–19.

[24] Elisabeth Hildt, "Artificial Intelligence: Does Consciousness Matter?" *Frontiers in Psychology* 10 (2019): 1535, https://doi.org/10.3389/fpsyg.2019.01535.

[25] Thomas Aquinas, *Summa Theologiae*, trans. Fathers of the English Dominican Province (Allen: Christian Classics, 1981). René Descartes, *Principles of Philosophy*, trans. V. R. Miller and R. P. Miller (Dordrecht: Springer, 1983).

[26] Nick Bostrom, Allan Dafoe, and Carrick Flynn, "Public Policy and Superintelligent AI: A Vector Field Approach," 2018, https://pdfs.semanticscholar.org/9601/74bf6c840bc036ca7c621e9cda20634a51ff.pdf (accessed October 7, 2019).

2.4 African Population Ethics

Comparatively little has been written on the African perspective of population ethics. As defined earlier, population ethics discusses the various ethical issues that arise when we consider how our actions affect both who is born and how many people are born (Greaves, 2017). We also underlined those African perspectives of population ethics that tend to look at the number of people born and not so much who is born, the circumstances of their birth, or how they will live thereafter.

I would like to borrow from Caesar Alimsinya Atuire (2022) when he writes about "African perspectives of moral status: a framework for evaluating global bioethical issues,"[27] though this topic is not directly about African population ethics. What is central to Atuire's paper is the basis for moral judgment from an African perspective, which, in this case, is the dual conceptions of human personhood. In his view, the moral status of a human person can be considered from both the objective and from the substantive points of view. Behrens (2011) already stated that, "Talk about moral persons can take two forms. A moral person can be seen either as an object (patient) of the actions or omissions of another or as a subject (agent) whose agency has a moral value" (quoted by Atuire 2022: 240). Atuire continues to state that:

This allows us to conceive of moral status in a dual manner. In the first place, we can talk of object moral status (OMS), where a being is worthy of moral consideration or is owed moral duties by an agent. We can also talk about subject moral status (SMS), whereby an agent's moral worth is linked to the moral quality of her agency. The African notion of full moral status (FMS) requires both OMS and SMS.[28]

Without going into the details of how Atuire's view on OMS is similar to Kantian deontology, which Atuire (2022, 240) ably discusses. The point that is central here is our interest in the moral judgment from an African perspective. The African perspective of population ethics is centered on these dual understandings of human personhood: person as an object of moral respect, and person as an agent or subject of moral actions. In line with population ethics, Africans will judge the worth of the human population from these two perspectives and not their welfare or well-being alone. This is because a human person who is born or who is not born yet, is worthy of living as long as they are an object of respect and an agent of moral actions.

What I think may need to be added to Atuire's contribution is what I consider about a human person, who is both an individual and a corporate person. Though Atuire links OMS to the human corporate existence and SMS with individual existence, our view differs in that humans are not just mere individuals who deserve moral respect and are subjects or agents of moral activities. My view is that this OMS should be qualified as one's relationship with self, others, and the

[27] Charles A. Atuire, "African Perspectives of Moral Status: A Framework for Evaluating Global Bioethical Issues," *Medical Humanities* 48 (2022): 238–245, https://doi.org/10.1136/medhum-2021-012229

[28] Atuire, African perspectives of moral status, 240.

environment, which includes the invisible world of the gods. It is this communal basis of human personhood that distinguishes it from the Western perspective.

Atuire summarizes this as follows:

First, OMS is grounded in being a member of the human species irrespective of capacities and agency. All that is required is to possess the metaphysical attributes of being human. These attributes are constitutively relational. Second, SMS is accorded in varying degrees to individual beings, humans and non-humans, based on their perceived agency in contributing towards communal humanism. We may say that while OMS has a holistic grounding, SMS has an individualistic grounding. FMS [Full Moral Status] is accorded to beings that have both OMS and a perceived high level of SMS.[29]

While the above view is correct, the basis of African population ethics is not about human relationships with the rest of humanity, which Atuire describes as holistic, or the value accorded to an individual as a contributing agent to communal humanism. An individual as a human person is both an individual, whose individuality is not determined by the rest of humanity, and a corporate being, which is metaphysical.

From the African population's ethical perspective, our view is that, on the basis of these African metaphysics, an individual is ontologically grounded in others, nature, and God, which makes an individual a unique person capable of finding meaning in human existence. The meaning of human existence will not be derived only from the level of their welfare or well-being, but mainly from the quality of their relationship with self, one another, the environment, and the spiritual world.

This view is partly supported by Thaddeus Metz, who, in his book *A Relational Moral Theory: African Ethics in and beyond the Continent* (2022), insists on relational moral theory. While Western population ethics seeks to judge the number of humans to be born based on how their welfare and well-being will be guaranteed, African population ethics judges who and how many people should be born by making reference to human relationships with one another, nature, and the spiritual world.

Metz (2022) distinguishes three different ethical theories in African ethics: African vitality ethics, which he associates with, among others, Noah Dzobo; African welfare ethics, which he associates with the work of Kwame Gyekye; and the relational approach, which he proposes. Since Part I of his book is on meta-ethics, Part II on normative ethics, and Part III on applied ethics, I take interest in his view on normative ethics. In this part of the book, Metz discusses (i) the common good (well-being); (ii) vital force (life); and (iii) communal relationship (harmony) theories.[30] Ochieng'-Odhiambo, in his review of this book of Metz, reiterates that: "The author [Metz] goes on to argue that it is the communal relationship principle which is promising as an independent foundation for right and wrong and not

[29] Atuire "African perspectives of moral status", 242.

[30] Thaddeus Metz, *A Relational Moral Theory: African Ethics in and beyond the Continent* (Oxford: Oxford University Press, 2022).

the first two."[31] Metz disagrees completely with the common good [welfare] theory proposed by Kwame Gyekye, but does agrees in part with the views of the proponents of the vital force theory.

Metz sees the vital force theory as being supported by thinkers like Noah Dzobo, Bénézet Bujo, and Laurenti Magesa. While not rejecting their views in totality, Metz argues that their views are unable to account for a variety of comparatively uncontroversial claims about ethics. For example, they would not be able to account for the wrongness in the principle of "separate but equal is still unequal" or the wrongness of telling innocent people lies or tricking them, even when there is little to no risk of impairing their lives.[32]

What is more important is what Metz says about the common relationship principle. He begins by underlining the fact that African tradition is "communitarian." What this means is that relational concepts of community, love, and harmony are prominent. He continues to argue that the notion of relationality implies one identifying with others. In other words, identifying with others is a major part of "harmony" or "communal relationship."[33]

Where I disagree with Metz is that he dissociates African relational theory from metaphysics by referring to it as purely a secular moral principle. By including God or the gods in this relationship, Metz thinks, it then becomes metaphysical. The fact is that the African relational theory of communion is incomplete if it includes only relationships among humans and between humans and their environment. This is because this human environment is regarded as both material and spiritual. Within their environment, besides the material dimension, there is the spiritual dimension, where the ancestors and their gods play significant roles. This cannot be excluded if we are to understand the basis of African ethics.

On the grounds of Western population ethics, unfortunately, a number of policies affecting the human population have been designed to systematically reduce the human population, especially where such a propensity to produce is higher. The continent of Africa is a victim of one such policy, officially showing a kind of sympathy or empathy with Africa's poor financial and economic status, thereby affecting their welfare. Seriously speaking, however, there is no sympathy or empathy with Africa. In their view, Africa's population increase is a danger to the Western trend, where there is a continuous decrease in population. The real problem is not about the welfare of the people of the world, but the welfare of the richer populations of this world in general, and of developed countries in particular.

---

[31] Frederick Ochieng'-Odhiambo, "Book Review on 'A Relational Moral Theory: African Ethics in and beyond the Continent'," *Philosophical Papers* 51, no. 3 (2022): 477–481.

[32] Frederick Ochieng'-Odhiambo, "Book Review on 'A Relational Moral Theory: African Ethics in and beyond the Continent'," *Philosophical Papers* 51, no. 3 (2022): 477–481.

[33] Thaddeus Metz, *A Relational Moral Theory: African Ethics in and beyond the Continent* (Oxford: Oxford University Press, 2022), 478.

2.5 Conclusion

I conclude by saying that, though a number of Western thinkers seem to show sympathy toward the population explosion trends in Africa, in reality, they are not sympathetic. They are only trying to be diplomatic about the situation. The naked fact is that they disapprove of the enormous upsurge in the number of people being born in Africa, yet their propensity to live a better quality of life is never guaranteed.

Secondly, value judgments in population ethics come from the wrong principle: the suggestion that the cause of poor quality of life in the developing world is the question of overpopulation alone. None of them wants to accept that the situation in a number of these developing countries is a result of colonialism and the modern globalization system. The modern state of globalization encourages two things, consumerism and unrealistic competition, which keep developing countries at the status quo.

Consumerism is an economic theory and a cultural phenomenon which emphasizes the idea that the increasing consumption of goods and services is a desirable goal, and that happiness and well-being depend on acquiring material possessions. Relational harmonious living with one another, with the environment, and the spiritual world, as emphasized by African ethics, is totally ignored. Consumerism is, then, the creation of excessive desires to acquire as many goods of this world as possible without any due consideration of the effects of such practices on future generations and the income of the consuming communities. Today, in the globalized world, there is the capacity to produce more goods than the world population needs. However, this is often not done in order to keep the prices of goods high just to maintain the rich people's high standards of living..

On the other hand, while competitions among the developed countries and the socio-economic institutions within them are between those with similar potentials to compete, in developing countries, the slogan is "willing seller, willing buyer." This principle knows very well that in developed countries, it is the corporate institutions that compete among themselves, some of them with financial capacities bigger than that of several poor countries combined, yet these poor countries and the poor individuals within them compete with the multinationals. Such unrealistic competitions go on unabated. For instance, in Africa, an individual producer of raw material needs to compete and negotiate with the industrialized countries, and the powerful multinational companies within them. In what way can that individual have the power to bargain? In answer to this question, surely, an individual does not have the power to bargain. This shows that the current imbalance of economic power between developed and developing countries is designed by the developed world, and it is meant to last forever.

Most developing countries are told to produce raw materials, whose markets are in developed countries. After processing these raw materials, these goods are then sold back to developing countries at prices they cannot afford. How ethical is Adam Smith's economic trade arrangement? The answer is that the situation, which permits a "survival of the fittest" type of competition, is not really ethical at all in the view of this chapter. Individuals in developed countries are protected by their social security systems, while individuals in developing countries have to battle for survival in this toxic world of unfair competition and economic policies. Yet these developing countries are then considered "poor" and are told their poverty is a problem for the world.

Lastly, we conclude that African population ethics offers the best solution for judging practices and policies in population control discourse. The world has to open up to the thinking that human well-being is not the sole measure of happiness in this world. Happiness in this world is determined by a set of meaningful relationships, including the sharing of material wealth and poverty among humans, respect for our material and spiritual environment. In addition, utilitarianism or consequentialism is insufficient to be used as the basis for judging who should be born and how many people should be born in this world, because African metaphysics has a better alternative.

References

Adam, Smith. 1776. *An inquiry into the nature and causes of the wealth of Nations*. http://www.ifaarchive.com/pdf/smith.

Aquinas, Thomas. 1981. *Summa Theologiae*. Translated by the Fathers of the English Dominican Province. Allen: Christian Classics.

Atuire, C. A. 2022. African perspectives of moral status: A framework for evaluating global bioethical issues. *Medical Humanities* 48:238–245. https://doi.org/10.1136/medhum-2021-012229.

Bentham, Jeremy. 1970. In *An introduction to the principles of morals and legislation*, ed. J. H. Burns and H. L. A. Hart. London: The Athlone Press.

Bostrom, Nick, Allan Dafoe, and Carrick Flynn. 2018. *Public policy and superintelligent AI: A vector field approach*. https://pdfs.semanticscholar.org/9601/74bf6c840bc036ca7c621e9cda20634a51ff.pdf.

Chomanski, Bartosz. 2019. What's wrong with designing people to serve? *Ethical Theory and Moral Practice* 22:993–1015. https://doi.org/10.1007/s10677-019-10029-3.

Coole, Diana. 2021. The toxification of population discourse: A genealogical study. *The Journal of Development Studies* 57 (9): 1454–1469. https://doi.org/10.1080/00220388.2021.1915479.

de Vries, Josef. 1972. Life. In *Dictionary of philosophy*, ed. Walter Brugger. Spokane, WA: Gonzaga University Press.Translated and edited by Kenneth Baker

Descartes, René. 1983. *Principles of philosophy*. Translated by V. R. Miller and R. P. Miller. Dordrecht: Springer.

Ehrlich, Paul R. 1968. *The population bomb*. 12th ed. New York: Ballantine Books.

GFN (Global Footprint Network). 2014. *Lighthouse case study*. New York: Global Solution Network.

Gustaf, Arrhenius, Krister Bykvist, Tim Campbell, and Elizabeth Finneron-Burns, eds. 2022. *The oxford handbook of population ethics*. Oxford: Oxford University Press.

Gyekye, Kwame. 2002. *African cultural values: An introduction*. Legon, Ghana: Sankofa Publishing Company.

ten Have, Henk, and Maria do Céu Patrão Neves. 2021. Population ethics. In *Dictionary of global bioethics*. Cham: Springer. https://doi.org/10.1007/978-3-030-54161-3_410.

Hildt, Elisabeth. 2019. Artificial intelligence: Does consciousness matter? *Frontiers in Psychology* 10:1535. https://doi.org/10.3389/fpsyg.2019.01535.

Hicks, John R. 1981. *Wealth and welfare*. In *of collected essays in economic theory*, vol. 1. Oxford: Basil Blackwell.

Lewis, R. Peter, and Stefan Sarkadi. 2023. *Reflective artificial intelligence*. pp. 1–19.

Maciej, Musiał. 2022. Can we design artificial persons without being manipulative? *AI & Society*. https://doi.org/10.1007/s00146-022-01575-z.

Malthus, Thomas Robert. 1966 [1798]. *First essay on population 1798*. London: Macmillan.

Metz, Thaddeus. 2022. *A relational moral theory: African ethics in and beyond the continent*. Oxford: Oxford University Press.

Ochieng-Odhiambo, Frederick. 2022. Book review on 'a relational moral theory: African ethics in and beyond the continent'. *Philosophical Papers* 51 (3): 477–481.

Oslo University of Oslo. n.d. Deadweight. https://www.uio.no/studier/emner/sv/oekonomi/ECON1220/h04/undervisningsmateriale/deadweight.pdf.

Searle, John R. 1980. Minds, brains, and programs. *Behavioral and Brain Sciences* 3 (3): 417–424.

Sen, Amartya. 2009. *The idea of justice*. Cambridge, MA: Belknap Press of Harvard University Press.

Slesnick, T. Daniel. 1998. Empirical approaches to the measurement of welfare. *Journal of Economic Literature, American Economic Association* 36 (4): 2108–2165.

Chapter 3
Rethinking Traditional Akan Proverbs on Fertility in the Context of Modern Population Challenges

Beatrice Okyere-Manu

Abstract Population growth remains a topical interdisciplinary project, particularly within the disciplines of development, environmental, economic, and population studies. This is due to the unavoidable threat it poses to humanity and the environment on which human life depends. Despite the availability of family planning programs, high fertility rates persist in most African countries, particularly among rural and poorer communities. This chapter explores the role of traditional Akan proverbs in challenging and shaping attitudes toward family size and population growth in Ghana. A qualitative analysis of selected Akan proverbs was conducted to identify themes related to fertility and family size. While there are proverbs that encourage smaller families, cultural emphasis on large numbers often prevails, influencing population growth patterns. Through the lens of the normative ethical theory of consequentialism, the study concludes by examining the implications of the findings and their potential contribution to the discourse on Africa's population growth. Understanding the cultural values embedded in proverbs can inform the development of more effective family planning programs and policies that align with traditional beliefs.

Keywords African community · Population growth · African beliefs · Cultural values · Population Council

3.1 Introduction

This chapter is informed by a visit to a number of traditional marriage ceremonies in Ghana. In all the celebrations I have attended, there was a point in the day when some family members and the elderly people present were called on to offer advice

B. Okyere-Manu (✉)
Department of Philosophy and Ethics Studies, University of KwaZulu-Natal, Pietermaritzburg, KwaZulu-Natal, South Africa
e-mail: okyere-manv@ukzn.ac.za

B. Okyere-Manu, P. Mwipikeni (eds.), *An African Approach to Population Ethics*, Advancing Global Bioethics 23,
https://doi.org/10.1007/978-3-032-24464-2_3

to the new couple. It was interesting to note the content of the pieces of advice that are normally given. Almost all of the advisers suggest that the couples give birth to more children and, where possible, to sets of twins. Sometimes, even the ministers blessing the marriages would pray for more children. In a few of the more traditional occasions, libations have been poured into evoking the ancestors to bless the couples with more children. Most of the suggestions were made by using proverbs. Among the numerous pieces of advice given through proverbs to couples was the following: "*Awododo ye*," which literally boils down to "More children are good."

I have always wondered why all of the advice encouraged the production of large numbers of children. On no occasion were the couples encouraged to have fewer children than they could handle.

In one of the ceremonies, an auntie of the groom, who was about 70 years old, emphasized the need to have more children in order to maintain the family name. She went on to state, "Don't worry about how you will bring them up because it is God who looks after children, so don't worry about how you will raise them up. God himself will provide." It is reflecting on these and many more similar sayings that prompted this chapter.

Adopting an ethical perspective, this chapter examines some selected traditional proverbs of the Akan community in Ghana regarding family and childbearing. The chapter is necessary because, to date, little scholarship has been available to unpack the role that proverbs and traditional beliefs on fertility play in the face of modern population challenges. Moreover, most scholarship in the area of development studies has failed to engage in traditional values and beliefs and their implications for development and population change as a product of values.

The chapter argues that in most African communities, particularly among the Akans in Ghana, even though there is an intricate interplay of proverbs that encourages smaller families, people tend to focus more on those that advocate high fertility. In these communities, the production of large numbers of children has been perceived as a source of strength, wealth, future investment, and honor for families.[1] However, the effects of rapid population growth, including the devastating environmental consequences for both current and future generations, have not yet been adequately addressed at the family, community, traditional leader, and governmental levels. The world's population reached 7 billion just after 2010, and it is expected to rise to about 9 billion by 2045. The population of Africa continues to grow rapidly, despite the availability of family planning programs. Even where government priority exists, the implementation of such programs often lacks the cultural resonance required to significantly alter fertility trends.

The lack of urgency may have many reasons; it is against this backdrop that this chapter adds to the ongoing discussions and debate. It argues that one significant key determinant of this trend is the deeply embedded religio-cultural perception and understanding of population and fertility, as articulated in many traditional African

[1] Okyere-Manu, Beatrice. (2016). Overpopulation and the Lifeboat Metaphor: A Critique from an African Worldview *International Studies in the Philosophy of Science* 30(3), 279–289. 287.

proverbs, songs, and dirges. These beliefs, practices, and values are deeply ingrained in the core identity of the African people and should not be overlooked in any discourse around population and development. Therefore, as argued by Okyere-Manu (2016) elsewhere, any proposed strategy to curb the perceived population growth in Africa must consider the people's cultural values, beliefs, and practices.

This chapter, therefore, seeks to analytically explore selected traditional proverbs of the Akan community in Ghana regarding family and childbearing, and their implications for the social and environmental consequences of the current population growth crisis from an ethical perspective. It further argues that a critique of proverbs is central to any discussion on contextual and socially relevant population ethics. Such a discussion must be relevant to the daily lives of ordinary Akan individuals in Ghana. Any population ethics discussion has to seriously consider the influence of proverbs on the lived experiences of the Akans.

Through the lens of the ethical theory of consequentialism, the chapter argues that the literal meaning of these traditional proverbs is not particularly helpful in the current discourse on population growth. Practically, what is required is a radical shift in our religious and cultural consciousness concerning large numbers. A shift in our cultural understanding is a sustainable approach to addressing the current population growth crisis, as it is likely to provide a means of breaking and altering habits through deliberate and ethical reflection. Stakeholders, including family and community leaders, traditional chiefs, religious leaders, and the government, can join forces to achieve this goal.

The rest of this chapter is divided into four sections. First, it outlines the extent of population growth and its impact on Ghana. Second, it discusses the importance of proverbs in the lives of the people in the country, focusing in particular on the Akan people. Third, the chapter presents a selected proverb and its literal and philosophical meanings. Finally, it draws out key conclusions from the discussion for policymakers and stakeholders.

3.2 The Extent of Population Growth in Ghana

Like most countries in sub-Saharan Africa, Ghana has experienced a rapid rise in its population since the early 1950s. From a population of about just 10 million in 1980, the country's current estimated population is 29.6 million.[2] The annual growth rate is 2.5%, suggesting that, if not controlled, the country's population is likely to double in under 30 years. The growth is expected to pose a threat not only to economic development but also to public health, employment, sanitation, the environment, food security, and the overall security of its inhabitants.[3] It is expected that

[2] Suleiman Mustapha (2018). Don't go beyond 3 babies: population council warns *Daily graphic*. 25th July, 16.

[3] Suleiman Mustapha. Don't go beyond 3 babies, 16.

continual developmental deficits will lead to problems such as corruption, pollution, and other environmental issues, as well as internal and external migration, and also social vices, including criminal activities like prostitution, drug trafficking, and the like.[4] As noted by the current chairperson of the Population Control Council, Dr. Letitia Adelaide Appiah, teenage pregnancies, child marriages, and youth unemployment have exploded in the country as a result of population increase. She noted that the rate of teenage pregnancy currently stands as high as 14% and child marriage at 27%.[5].

Nevertheless, we are aware that the size of a country's population has many potential benefits to economic development, and Ghana is no exception to this. The benefits include, and are not limited to, the following: sections of the workforce in the country associated with both low and high skill levels, as well as the amount of goods and services that all consumers in the country are willing and able to acquire at a particular price: "the ability to consume and cause an increase in demand and the growth of GDP, its geopolitical importance, its tax revenue potential and the potential among others. It also has the potential to generate competition, efficiency, and innovation."[6] Despite the benefits, the chapter argues that a balance must be struck in order to safeguard the better living conditions of the Ghanaian people and future generations.

For this reason, in recent times, the Ghana Population Council has been pushing for the government to enforce a policy to put a 'cap' on the number of children one can produce to control the ever-rising population. The aim of this policy is to reduce the number of children a woman is likely to have in her lifetime and during her reproductive years to an average of just three.[7] The council announced that, at present, the population is growing in an uncontrolled way because some couples have more than ten children. It further argued that the government should review its free maternal policy; otherwise, the population would likely double at the speed mentioned above. The chairperson of the Population Control Council once again warned that there will be stringent fines for offenders.[8] This announcement sparked public debates, discussions, and disagreements. Whereas many have embraced this suggestion, a number of people, both literate and illiterate, have expressed their discontent, raising issues such as the preservation of cultural practices like polygamy, the lack of capacity to enforce such policies, and the rights of the individual couple, to mention a few.[9]

Such responses make this chapter a crucial contribution to the broader debate, as it challenges and critiques these beliefs and practices.

---

[4] Suleman Mustapha. Don't go beyond 3 babies, 16.

[5] Norvan Acquah Hayford. Business and Financial Times online (B&ft) Chat 2018 https://www.youtube.com/watch?v=KmKNYj-BUBY. Access date: 23 August 2018.

[6] Emmanuel Yao Anudjo. (2015). The Population Growth - Economic Growth Nexus: New Evidence from Ghana. *An Unpublished M.Phil. thesis*: University of Ghana Legon.

[7] Suleiman Mustapha. Don't go beyond 3 babies, 16.

[8] Suleiman Mustapha. Don't go beyond 3 babies, 16.

[9] Oswald K Azuma. 2018. https://www.myjoyonline.com/news/2018/July-25th/population-expert-rubbishes-npc-proposal-for-3-children-policy.php. Accessed date: 20 August 2012

3.3 The Importance of Proverbs in Ghanaian Traditional Society

Oral literature has been part and parcel of the folklore of African communities. It embodies history, cultural values, philosophy, and beliefs. In reality, it teaches the right things a society must do and the wrong things they need to abstain from.[10] It is comprised of riddles, puns, tongue twisters, proverbs, recitations, chants, songs, and dirges, which were created by traditional African peoples as part of their language and handed down from generation to generation.[11] Proverbs form a larger part of traditional African cultures; they inform all aspects of social life. John Mbiti, one of the prominent African philosophers, has argued that proverbs are deeply rooted in African culture. To him, anyone who has lived in any rural African village before is known to be a living carrier of proverbs.[12] Rulers and elders in rural African communities are the primary gatekeepers of proverbs, and their setting allows them to easily transfer these important values to their young ones. John Pobee reiterated that "proverbs are a vital and important mode of communicating and key to penetrating the worldview of Africans."[13] They are often short sentences that contain the deeper meaning of life and situations. It is believed that these short sentences are drawn from experiences to sum up situations in a metaphorical way, and it is a means of transferring information, values, and a code of conduct.[14] It is also seen as a source of moral and political guidance.[15] In effect, proverbs are seen "as compact expressions of important cultural knowledge, proverbs combine a cognitive economy of reasoning with pragmatic force aimed at influencing other people."[16]

In describing the role of proverbs among the Akan communities in Ghana, Nketia popular Africa's premier musicologist, remarks:

The value of the proverb to us in modern Ghana does not lie only in what it reveals of the thoughts of the past. For the poet today, or indeed for the speaker who is some sort of artist in the use of words, the proverb is a model of compressed or forceful language. In addition to drawing on it for its words of wisdom, therefore, he takes an interest in its verbal techniques—its selection of words, its use of

---

[10] Mirambo, I., Oral Literature of the Sukuma. P113–122. http://www.folklore/vol126mirambo.pdf. Accessed date: 20 August 2018, 121

[11] Chesaina, C. 1994, Oral Literature of the Kalenjin. East Africa: Educational Publishers Ltd. Nairobi.: viii).

[12] John Mbiti. 1997. The ethical nature of God in African religion as expressed in African proverbs, in Saayman, W (ed.), *Embracing the baobab tree: The African proverb in the twenty-first century*, 139–162.

[13] John S. Pobee. 1979. Towards an African Theology (Nashville, TN: Abingdon).

[14] Emmanuel Kweku Tetteh. 2013. Literary features of Akan Proverbs. Accra: Dwumfour.

[15] Gyekye, K.(1997). Philosophy and Human Affairs. New York: Oxford University Press; Quarcoo, A. K. (1972). The Language of Adinkra Patterns. Institute of African Studies, University of Ghana: Legon

[16] Holland, D., and N. Quinn. 1987. Cultural Models in Language and Thought: Cambridge University Press

comparison as a method of statement, and so on. Familiarity with its techniques enables him to create, as it were, his own proverbs. This enables him to avoid hackneyed expressions and give a certain amount of freshness to his speech. This … approach to proverbs which is evident in the speech of people who are regarded as accomplished speakers or poets of a sort makes the proverbs not only a body of short statements built up over the years and which reflect the thought and insight of Ghanaians into problems of life, but also a technique of verbal expression, which is greatly appreciated by the Ghanaian. It is no wonder, therefore, that the use of proverbs has continued to be a living tradition in Ghana.[17]

Thus, proverbs serve an important role in the oral history of the Ghanaian communities, and they are used not only by the older generation in their daily communication but also to exert authority and transfer truths. Proverbs often use figures of speech or metaphors to transmit their messages. No wonder it is perceived as an integral part of the African culture. It is important to note from the ongoing discussion that proverbs are used as figures of speech to transfer wisdom, values, knowledge, and history to the younger generation. While some are used as warnings, others provide suggestions for a better life.

Salwa Ahmed also has this to say:

Traditional proverbs have stemmed from various sources, often connected with individual experiences, observations, or general facts from different surroundings. People then refer to them on similar occasions and transmit them from generation to generation until they become part of their heritage. This is to say, a proverb is first uttered by an individual, and in time it becomes part of the traditions. People epitomize their way of thinking and ideas through proverbs.[18]

The above shows that proverbs are respected and perceived as an essential embodiment of the experiences of the ancestors that need to be carried on from generation to generation as a form of instruction. Pachocinski has observed the importance of proverbs and says that it is "still used in the daily lives of Black Africans, it flows from their blood, because it is through proverbs that traditions, history, wisdom, and ethics are transmitted from generation to generation to prevent it from dying."[19] Indeed, they are perceived as wise sayings that have won the acceptance of our ancestors for a long time and, therefore, must win the approval of modern society and be circulated and preserved. It is often used to encourage members of the traditional culture to perform their daily roles and duties with diligence. Within the informal setting of a home or within the family, it is often used to rebuke, advise, and instruct acceptable behaviors and warn against unacceptable ones. Quan Baffour draws our attention to the fact that proverbs among the Akans are still relevant, even in the twenty-first century because of their numerous functions. He explains that proverbs have educational significance since they have remained

---

[17] Joseph Hanson Kwabena Nketia. 1958. Drum Proverbs. Accra: Voices of Ghana. 21.

[18] Salwa Ahmed. 2005. Educational and Social values expressed by Proverbs in Sudan and England. Unpublished PhD Dissertation: technical University Berlin. 25.

[19] Pachocinski, R. 1996. Proverbs of Africa: Human Nature in the Nigerian Oral Tradition; An exposition and Analysis of 2600 Proverbs from 64 people. USA: Professors World peace academy. 1

essential to Ghanaian culture, heritage, and identity. It continues to serve as an educational tool around issues of birth, childhood, human origin, death, earth, universe, or *Onyame* (God and the Supreme Being).[20] It is often used extensively in formal gatherings and social functions such as durbar, conflict resolution processes, funerals, marriage ceremonies, and most initiation ceremonies to instruct, honor, and advise people. In short, African proverbs describe and prescribe the right behavior of people in the Akan traditional society.

Despite the importance of proverbs in Akan traditional society, it has been observed that most proverbs are slowly losing their usage and relevance due to the current transformation of Ghanaian societies. However, Mieder has a different opinion, and he has argued that:

Proverbs are flexible and adaptable to every new context and interpretation because their metaphorical language is not limited to a specific context. Changing times require forms of expression that the traditional proverbs can no longer supply. However, it often suffices to adapt and apply proverbs to the modern context.[21]

I agree with Mieder that although most African communities extensively use proverbs as part of their cultural heritage, given the reality of modernization, globalization, and the interaction of cultures through migration, these proverbs may require analytical discussion. This is crucial since culture is not static but dynamic. It is important to change and adapt with time. An Akan proverb says, "*Mere dane a dane wo ho bi.*" This proverb literally means "When times change, one needs to change with it."[22] This implies that change is inevitable in life. At any time, situations can change for good and sometimes not so good; whenever there is change, there has to be an effort to adapt to ensure maximum survival. Relating this understanding to the ongoing discussion suggests that some of the proverbs, dirges, and songs may require new forms of expression and emphasis because the traditional form of expression may be unable to supply suitable meaning to our current situations. For this reason, there is a need to interrogate some of the old proverbs that speak to issues around population.

From the above discussions, the chapter has highlighted that a major contributing factor to the exponential growth of the Ghanaian population has been the influence of our traditional beliefs and practices embedded in our proverbs. Until reasonably recently, fertility issues have always been a private issue among couples and the elders in families. The adverse effects of high fertility on society, individual citizens, and the environment, as pointed out above, have forced these issues into the public domain. It must be noted that the government and policymakers have not been well-equipped to deal with questions surrounding such private issues as fertility, as proposed by our proverbs. How does one decide for someone else how many

---

[20] Quan Baffour. The wisdom of our forefathers: Akan Proverbs and their contemporary educational values. *Southern African journal of Folklore studies* Vol 21 (1), 2011

[21] Mieder Wolfgang. (1993). *Proverbs are never out of season.* New York: Oxford University Press, xi

[22] Yankah, K. (1992). Traditional lore in population communication: the case of the Akan in Ghana. *Africa Media Review,* 6(1), 15–25.

children they must have in this day and age? Yet high fertility comes with a number of ethical questions. Why will someone decide to have children when they know they cannot look after them? Why should taxpayers be forced to pay large sums of money to cater to people who do not plan their families? Why should someone give birth to children they know they cannot look after? Why should children be made to stop school in order to look after their siblings or sell to support their parents because they are poor but have large families? Why should children be made to sacrifice their future? How can children realize their potential in such situations?

In an attempt to help address these questions, the government has faced a negative reaction from the masses to the with Ghana Population Control Council (GPCC) suggestion. A critical examination of the responses reveals the deeply ingrained nature of our cultural beliefs. An appropriate response to this crisis is to actively work toward interrogating these embedded proverbs and why they were constructed. Such discussions will help contextualize the proverbs to see their relevance in modern society, and. as Yankah has rightly put it, "traditional modes of communication [oral literature] may change themes or style or may adapt themselves to novel situations as the need arises."[23]

3.4 Discussions of Akan Proverbs and their Relations to Fertility Patterns

This section of the chapter discusses selected Akan proverbs related to fertility. It will examine the literal and philosophical meanings of these terms and their relevance in the current discourse on population growth. The philosophical meaning of these proverbs is that the Akans believe in the community. As a communitarian society, each individual depends on others within the community; there cannot be a community with just a few individuals. This underlines the need for more people. It must be noted that most of the proverbs below have multiple uses:

"*Awo dodow ye.*" This proverb literally means "Many children are good."

As noted above, the Akan adhere to a communitarian worldview, believing that individuals are integral to the community. Therefore, to maintain families and the community, it is prudent for couples to have more children.

This proverb has multiple uses. It is used to encourage people to have more children. It is also used when children do something that honors and brings happiness to their parents. "*Abusua ye dom*" is another proverb, which literally means "Family is a multitude." The key understanding of this proverb is that the lineage/family is not a few members, but a crowd. Succession is an important aspect of Akan traditional life, and, therefore, they encourage family members to have more children to

[23]Yankah, K. (1992). Traditional lore in population communication: the case of the Akan in Ghana. *Africa Media Review,* 6(1), 15–25. (23)

continue the family line. To them, there cannot be a family or ancestry with only a few individuals. When disaster strikes, members will be there to share the burden and encourage one another.

Another important proverb in this category is that *the Cotton Threads say they need to be in a large company to be able to carry a stone*. This suggests that a cotton thread on its own is fragile and cannot have its seeds, but when there are many, they are able to carry their seeds on the cotton plant. This is an Ewe[24] proverb. Its broader meaning is that burdens carried by a number of people become lighter; therefore, many children can help share the problems within the family. This proverb has multiple uses; it is either used to advise prospective couples or, in most cases, when settling family and community disputes.

Akoko takra na ekyere se akoko no ye kese or anomaa ahooden ne ne ntakra.

The literal meaning of this proverb suggests that the number of the hen's feathers enhances its size; it can also indicate that the health of the hen is determined by its feathers, as the bird derives strength from its numerous feathers. This is an important proverb in the Akan tradition because the hen is regarded as a vital asset. It produces eggs and offspring (chickens), and it has economic and religious usefulness. A well-fed hen or cock symbolizes wealth. Therefore, a person's possession (particularly his children) determines his standing in society. It is used as a praise for great people with numerous children who have made it in life.

Yede nnipa na ekyekyere kuro. This is another proverb that advises couples who want to start their own families. It literally means that one cannot build a town or a city with only a few inhabitants, signifying the need for a larger population. "We use multitudes to establish a town/city." This implies that they cannot rely solely on a few children to build their own legacy, let alone that of the family. Multitudes were important to the Akans, not only to extend the lineage into the future but also in times of war.

Like most Akan proverbs, it has multiple uses; it can also be used to settle disputes among family members or within the community, encouraging unity and peace.

Nsamanfo mpo pe won dodoo, na enye ateasefo literally means "Even the dead love numbers. That is why they keep adding onto their numbers and how much more they are living."[25] This proverb encourages the living to increase their numbers or accept more people in their communities, as even the dead desire to be in the company of many. Therefore, they continue to invite people daily to join them. The analogy to the dead here is very interesting because the Akans believe death is not the end of life. There is life after death; if the dead want more people to join them, the living should encourage more to do so. It also suggests that it is a blessing for the living to increase in numbers because there is virtue in numbers. This proverb is often used to advise couples who prefer having fewer children. It is also used as a warning in situations where individuals exhibit indifference toward each other.

---

[24] Ewe is an ethnic group in Ghana, found in Southern Ghana and Togo, Djifa Kothor. 2012. Home Away From Home: A Study of the Ewe Unification Movement In The United States. Unpublished master thesis. University of Illinois at Urbana-Champaign), 7.

[25] Yankah, Traditional lore in population communication, 19.

Dua baako nye kwae literally suggests that one tree cannot make a plantation. As farmers and hunters, the Akans understand the importance of forests as a resource for food, fuel, and soil improvement. In fact, the Akans believed that their lives depended on the forest or plantations; therefore, there was a need to maintain it. This proverb also has multiple uses; in most cases, however, it is used to advise couples with few children (or one child) of the importance of having more children. The plantation is used here to represent the family or community.

Nsatea baako butu fa ade wo fam a entumi. The literal meaning of this proverb is that if one figure tries to pick up something from the ground, it will not be able to do it. The same idea is conveyed in a similar proverb, "*Nsa baako nsoa adesoa,*" which means that a hand is never enough to lift a heavy load. The explanation of this proverb is that there is power and strength or unity in numbers. This proverb is used to encourage high fertility and maintain communal spirit. It resonates with the phrase *Baanu so a emmia*, which suggests that two people are able to share loads or burdens. A load shared makes it lighter. Similarly, it also implies that we are not created to live independently of others; we need other people at every stage of our lives. Thus, a couple need not become selfish and have only one child; no matter how much the child achieves when he/she works with others, they are able to do much more than they would have done on their own.

Traditionally, most Ghanaian communities have been made up of farmers who owned large stretches of land and substantial numbers of animals. Each community had large stretches of land, and the traditional leaders, namely the chiefs, were the overseers. One of their responsibilities was to ensure that the various community clans had access to land. The leaders of the society, in turn, allocated portions of the land to the family heads, whose responsibility was to ensure that each member of the family had a portion to cultivate for themselves and their families. During the clearing, tilling, planting, and harvesting seasons, it was difficult for the individuals to work on their land. Each one had their own piece of land; therefore, there was no one available to be a laborer. The farmers, therefore, depended on their families, particularly their children, for the cultivation of their land. In order to have enough children to help on the farms, polygamy was encouraged, which is why there were large families.

Again, one of the beliefs held in high esteem in most traditional, and even modern, Ghanaian communities is motherhood. Meyre Ivone da Silva (2004) reiterated Buah's position that Africans regard fruitfulness in marriage as having many children. It does not matter whether the family is wealthy or not. Buah has observed that women with many children are respected and perceived as blessed in their marriages, particularly when they have sons. The belief is that male children will eventually grow, marry, and work hard to look after their mothers and fathers in their old age. In a patrilineal setting, the son usually becomes the custodian of whatever inheritance the father may leave behind. They extend the family lineage, and their wives typically assume household chores and serve their mother-in-law and all family members. Therefore, the belief in these settings is that when a couple gives birth only to girls, they will continue to try for boys and eventually end up with lots of children. Unfruitful or barren women are considered unblessed in their marriage; the result is that they often end up divorced, even sometimes being perceived as "evil."[26] Peter Sarpong describes childlessness in the broader Ghanaian society as "disastrous." This is because, apart from being the leading cause of divorce, a child is one of the conditions for being deemed an ancestor. Sarpong believes that a key condition for being perceived as an adult in society is marriage. Thus, a 15-year-old married girl is considered an adult, whereas a 60-year-old unmarried man cannot be accorded the same status. He explains that "one of the conditions of one becoming an adult is determined by marital status and the number of children one has. Bachelorship suggests a refusal to increase the number of relatives one possesses; such a person is deemed useless and not considered an ancestor."[27]

When it comes to the issues surrounding high fertility, most Akans believe that in life, accidents and tragedies can occur; therefore, having a large number of children provides a safety net in the event of a disaster, such as a chronic disease or death, as there will still be more children to extend the lineage into the future. Again, Okyere-Manu reiterated Samuel K. Gaisie's (1972) claim that childbirth is perceived as a guarantor of financial stability.[28] The assumption, therefore, is that the more children one has, the more secure their future will be because at least one or two of the children will not only carry the responsibility of caring for the parents in their old age but will also look after the younger siblings, aunties, uncles, nephews, and nieces. Dzobo (1992) has argued that:

Historically, the emphasis upon having many children in Africa was not a sign of irresponsible parenthood. It was a way of understanding the essence of human existence, which involves full participation in humanity's creativity and the passing on of the creative power of life from one generation to another through the processes of marriage and procreation. A satisfactory discharge of this responsibility partly earns one the title of *Nana*.[29]

Extending Dzobo's argument, which is shared by many other African philosophers, it can be asserted that women's prestige is found in the number of children they have. Most Akan women are celebrated and rewarded with *Badudwan* for giving birth to a tenth child (a female rewards the husband with a ram upon giving birth to the tenth child). Not only are the living praised for having more children at funerals, but the dead are praised and honored through songs and dirges, even at their funerals. An example of such songs of praise includes:

Eno nkruma kese a ne yam abaduasa (Mother, the okro that is full of seeds. Or Eno the okro full of seeds that rears others' children as well). Also in the following line from a hunter's song (abofodwom), a hunter's virility receives mention,

---

[26] Meyre Ivone da SILVA. 2004. Revista de Letras, São Paulo, 44 (2) 2004: 129–138.

[27] Sarpong, Ghana in Retrospect, 34–35.

[28] Okyere-Manu, Overpopulation and the Lifeboat, 283.

[29] Dzobo, Noah Komla. (1992). Values in a Changing Society: Man Ancestors and God Person and Community: Ghanaian Philosophical Studies 1. Edited by Kwasi Wiredu and Kwame Gyekye. Washington: The council for Research in Values and Philosophy, 232.

Twenedwase Anokwa Tomfoo a ne mma refuafua no (Anokwa the smith of Twenedurase whose children are clinging to him).[30]

3.5 Rethinking Traditional Proverbs in Light of Ghana's Population Growth

The discussions to date suggest the need to rethink the literal meanings of traditional beliefs and practices in light of the current overpopulation debate. The pertinent question to ask is: Are these literal meanings helpful, considering the enormous challenges of population growth? I believe our ethical interrogation must empower us to reconsider some of the literal meanings and their negative effects. In doing so, it is essential to recognize and focus on the impact on individual couples, families, communities, and society as a whole. For instance, how can the number of children be perceived as a sign of wealth or honor, particularly for poor and struggling families in the midst of the current economic, developmental, and environmental crisis? This, along with the many unanswered questions posed earlier in the chapter, calls for the consequentialist approach to ethics as a lens through which to interrogate these proverbs. Martin Peterson has noted that the normative ethical theory of consequentialism suggests that the morality of an action can be judged by the consequences resulting from that particular action.[31] This implies that the consequences of an action determine whether the action is right or not. Relating this approach to the foregoing discussion suggests that the consequences of having a large family determine whether the choice of the number of children is right or wrong, particularly in a poor family. Through the lens of the theory of consequentialism, what seems to emerge from our discussion is that the consequences of the literal meaning of the proverbs suggest the following: a barrier to development; a burden to the parents in terms of affordability to struggling families; and a burden on the nation, community, and the family as a whole. The consequences also suggest a shortage of resources, environmental degradation, and an unsustainable future for the next generations.

In the midst of this, Yankah's observation becomes very important. He has drawn our attention to the fact that, in spite of the traditional emphasis on high numbers, "traditional lore drops hints of the caution African society's exercise against excesses for the socio-economic dislocation resulting from excessive numbers is not unknown in traditional Africa."[32] Yankah argues that, although Akans emphasize large numbers, other proverbs warn against having too many children. Interestingly, in most instances, such proverbs are not visibly high-

[30] Nketia, J. H. Kwabena. 1955. Funeral Dirges of the Akan People. Accra: Achimota Publishers, 62

[31] Martin Peterson. (2003). From Consequentialism to Utilitarianism. *The Journal of Philosophy*, 100(8), pp. 403–415. (403)

[32] Yankah, Traditional lore in Communication, 22.

lighted; they are often mentioned occasionally to warn against the socio-economic consequences of high fertility after people have had many children. An example of such a proverb includes *Prako awo ye musu*. The literal meaning here is that too many children are derogatorily likened to pigs, as in the saying, "Pig birth is a curse." In this instance, "pig birth" refers to the multiple births typical of pigs.[33] This proverb is sometimes used to describe various births with shorter birth spacing, along with its financial implications. A pig can give birth to between 7 and 12 at a time, with the piglets often weaned at 3 months of age. It is perceived as a curse due to the strain and struggles the mother pig endures until the weaning is complete, as she eats less and does not get enough rest and sleep. Quite apart from this proverb, there is a similar one that also warns against high fertility and its economic effects, which Yankah found in his study: *Awo dodoo nti na mannya dwetire*. The proverb refers to a woman whose capital for her business was spent on her children instead of investing it in her company. She has had to use the capital to feed and clothe the numerous children.[34] While emphasizing the role and significance of small numbers, Yankah again narrated a story to explain this proverb: "Many births and rags are bedfellows," which describes the tragic death of a woman named Ama Awusi, who had thirty children. In the narrative, he mentioned that they had a single snail; when all thirty children demanded their share, because she did not know how to divide the snail among that number, she was forced to swallow it, choked, and died instantly.[35] The point is that when there are many children in the family, unhealthy competition can arise, even in conditions of limited family resources.

Indeed, these warnings hold valuable lessons that must be highlighted to help balance the dependence of large families. However, one wonders why these proverbs have not received the same attention. One way to make sense of this tension is that most Akans assign moral significance to numbers, as indicated above, and overlook the negative impact accompanying it, as observed in the ceremony I attended. Interestingly, a guest advises the couple that God will look after the children, so they do not have to worry, but they should go ahead and have more children. This culminates in the saying "*Onyame na owe nipa*," suggesting that having more children is beneficial because God, the provider, will take care of them. The point here is that children are a gift from God, so He will surely provide for them. This implies a shift of responsibility from the parents and family to God, something which is not helpful in our current religiously pluralistic society. How do we expect God to physically look after the numerous vulnerable children in our communities? So, practically, what do we do? There is a need to revisit these proverbs and re-examine their meanings in light of the current challenges we face.

---

[33] Yankah Traditional lore in Communication, 22.

[34] Yankah, Traditional lore in Communication, 22.

[35] Yankah, Traditional lore in communication, 22.

3.6 Conclusion

This chapter discusses several Akan proverbs related to fertility. It has been argued that the high fertility rate among the Akan people is the result of the socio-cultural importance of families to them. Traditionally, they needed more people to help the farmlands. A high value was also placed on the male child, which encouraged couples with only girls to try repeatedly for a boy. The Akans believe that marriage and childbearing are a mark of adulthood and a condition of ancestorship after death, thereby fueling the desire to have more children. Again, a woman's prestige was found in the number of children they had. The chapter further argued that, although such beliefs and practices, which are enshrined in proverbs on high fertility, help to reinforce the cultural heritage of the Akan people, their literal meanings are not particularly helpful with regard to addressing the current population growth and environmental degradation crisis. It noted that though there are some proverbs that warn against high fertility, they are often not highlighted in the daily communications of the Akans. From the discussion, the chapter concludes that there is a need to emphasize proverbs that encourage low fertility so sa to balance the dependence on large families in meeting modern and future needs associated with overpopulation.

References

Acquah, Hayford Norvan. *Business and financial times online (B&ft) Chat 2018* https://www.youtube.com/watch?v=KmKNYj-BUBY. Access date: 23 August 2018.

Anudjo, Yao Emmanuel. 2015. *The population growth–economic growth nexus: New evidence from Ghana. An Unpublished MPhil. thesis:*. University of Ghana Legon.

Azuma, K. Oswald. 2018. https://www.myjoyonline.com/news/2018/July-25th/population-expert-rubbishes-npc-proposal-for-3-children-policy.php. Accessed date: 20 August 2012.

Beatrice Okyere-Manu (2016) Overpopulation and the Lifeboat Metaphor: A Critique from an African Worldview, International Studies in the Philosophy of Science, 30:3, 279-289, 281.

Chesaina, C. 1994. *Oral literature of the Kalenjin*. East Africa: Educational Publishers Ltd. Nairobi.

Da Silva Ivone, Meyre. 2004. *Revista de Letras*, São Paulo, 44 (2) 2004: 129-138,

Gyekye, K. 1997). Philosophy and human affairs. New York: Oxford University Press; Quarcoo, A. K. (1972). *The Language of Adinkra Patterns*. Legon: Institute of African Studies, University of Ghana.

Holland, D., and N. Quinn. 1987. *Cultural models in language and thought*. Cambridge University Press.

Komla Noah, Dzobo. (1992). Values in a changing society: Man ancestors and god person and community: Ghanaian philosophical studies 1. Edited by Kwasi Wiredu and Kwame Gyekye. Washington: The council for Research in Values and Philosophy, 232.

Mbiti, John. 1997. *The ethical nature of god in African religion as expressed in African proverbs*.

Mustapha, Suleiman, (2018). Don't go beyond 3 babies: Population council warns *Daily graphic*. 25th July, 16.

Mirambo I. *Oral literature of the Sukuma*. pp 113–122. http://www.folklore/vol126mirambo.pdf. Accessed 20 Aug 2018, 121.

Nketia, Hanson, and Kwabena Joseph. 1958. *Drum Proverbs*, 21. Accra: Voices of Ghana.
Nketia, J. H. Kwabena. 1955. *Funeral dirges of the Akan people*, 62. Accra: Achimota Publishers.
Okyere-Manu, Beatrice. (2016). Overpopulation and the lifeboat metaphor: A critique from an African worldview. *International Studies in the Philosophy of Science* 30(3), 279–289. 287.
Peterson, Martin. 2003. From consequentialism to utilitarianism. *The Journal of Philosophy* 100 (8): 403–415. (403).
Pobee, S. John. 1979. *Towards an African theology*. Nashville, TN: Abingdon.
Saayman, W (n.d.) *Embracing the baobab tree: The African proverb in the 21st century*, 139–162.
Tetteh, K. Emmanuel. 2013. *Literary features of Akan Proverbs*. Accra: Dwumfour.
Yankah, K. 1992. Traditional lore in population communication: The case of the Akan in Ghana. *Africa Media Review* 6 (1): 15–25.

Part II
Cultural and Philosophical Perspectives

Chapter 4
Meaning of Life and Population Issues from an African Perspective

Wilfred Lajul

Abstract Population discourse centers around the number of people living and being born into the world, which is not increasing in size. Imperatively, this discourse points to life and its meaning in the world. This is because the more lives are brought into the world, the stiffer the competition for the resources on which humans depend, thus undermining the basis for having a meaningful life in this world. Our question, however, relates to the manipulative devices humans are using to address this situation. This leads to the question of whether life is to be considered a commodity that humans can manipulate and shape at will, or a given entity to be cherished, nourished, and protected. Modern Western science, with its propensity to manipulate life, tends to cherish the former, while the African humanist philosophy takes the latter view, resulting in significant disagreement between the two approaches. Population ethics, on the other hand, examines the various ethical issues that arise when considering how our actions impact both who is born and also the number of people born. While Western philosophy emphasizes the human duty to determine who is born and how many people are to be born, African perspectives tend to value the number of people born rather than who is born, regardless of the circumstances in which they are born or how they will live after birth. Analysing these contrasting views, this chapter questions whether the meaning of life is determined by who is born and how many are born in this world, or whether these factors are inconsequential to establishing the meaning of life. Using a critical analytical method, the chapter digs deeper into the implications of these divergent views in the context of population discourse. The hope is that a deeper understanding of the question of life, which provides reasons for the meaning of life, will reveal that human life is actually more than what human science can determine. Although humans play a significant role in determining the meaning of life, there are factors beyond human beings to consider in establishing the meaning of life in relation to who is born and how many people are born in this world.

W. Lajul (✉)
Gulu University, Religious Studies and Philosophy, Faculty of Education and Humanities, Gulu, Uganda

B. Okyere-Manu, P. Mwipikeni (eds.), *An African Approach to Population Ethics*, Advancing Global Bioethics 23,
https://doi.org/10.1007/978-3-032-24464-2_4

Keywords African humanism · Population bomb · Food production · African metaphysics · Explosive population

4.1 Introduction

In 1968, Paul Ehrlich's book, *The Population Bomb*, famously declared that unless countries engage in population control, "hundreds of millions of people will starve to death in spite of any crash programs embarked upon now."[1] Ehrlich went on to observe that, "While there are temporary or local reversals of this trend, it now seems inevitable that it will continue to its logical conclusion: mass starvation."[2]

However, some 50 years after the predicted mass starvation of the people in the world, David Lam observed that the bomb had not yet exploded. He indicates that during this period, "… the world grew from 3 billion in 1960 to 7 billion in 2011, an unprecedented rate of population growth that will almost surely never be seen again, the world experienced some of the biggest improvements in standards of living in human history."[3]

Lam continues to show that "food production grew faster than population in every major region between 1961 and 2009. […] My interpretation of this history was that it was a period of surprising success, with the world adding 4 billion people while also improving standards of living in many of the world's poorest countries and avoiding major resource scarcities".[4]

These contrasting episodes illustrate the impact and magnitude of speculation and myths surrounding world population. The direct results of Ehrlich's "population bomb" catchphrase led to a focus on drastic measures to curb population growth using various methods. Though not unique to population control, these measures include the legalization of abortion in a number of countries in the world, the increased use of contraception, the introduction of genetically modified foods, and the reduction of population ethics to utilitarianism, complicating the moral question of life..

Ehrlich's speculation and myths surrounding the human population have raised a serious question about the meaning of life on Earth. On one hand, the meaning of life seems to be determined by its quality, supposedly brought about by controlling the size of the human population; on the other hand, there is a serious critique of such control. In this critique, the meaning of life appears to be influenced by factors which go beyond the singular population equilibrium proposed by Ehrlich in 1968.

---

[1] Paul R. Ehrlich, *The Population Bomb*, 12th ed. (New York: Ballantine Books, 1968, xi).

[2] Paul R. Ehrlich, *The Population Bomb*, 1

[3] David Lam, "Has the World Survived the Population Bomb? A Ten-Year Update," *Population and Environment* 45 (10). 2023, 1, https://doi.org/10.1007/s11111-023-00422-7

[4] David Lam, 2023, 1–2.

This chapter discusses the significance of this assumption on the meaning of life in the world. It raises the question of whether or not there is a direct relationship between the human population and the meaning and quality of life in this world. The hypothesis is that the meaning of life does not depend solely on the number of people living within a given environment, but on other equally important factors, such as the level of cooperation and sharing among humans, harmonious coexistence with nature, and recognition of God's plan, among others. Western philosophy, with its proclivity to relegate the role of God to the periphery and to manipulate humans and nature, tends to support the view that the quality and meaning of life are directly related to human population size. Contrastingly, African metaphysics, with its emphasis on the ontological coexistence among humans, between humans and nature, and between humans and their God, tends to support the view that there is no direct link between human population size and the quality and meaning of life in this world. In this chapter, however, none of these two positions will be taken for granted, because each will be critically examined in terms of its relevance in the contemporary world. Employing a critical analytical method, this chapter examines the concept of life in the context of population discourse, with a particular emphasis on African philosophical perspectives.

4.2 The Meaning of Life in Philosophical Thinking

Both Western and African philosophical views seem to sway between the views that humans are duty-bound to create meaning for their lives and the view that it takes more than human efforts to create meaning for human life, respectively. Aribia Attoe, in his edited book *African Perspectives on the Question of Life's Meaning*,[5] discusses the philosophical theories of the meaning of life in both Western and African philosophy. Beginning with Western philosophy, in that book, he identified three theories of meaning: supernaturalism, naturalism, and nihilism.

Supernaturalism emphasizes that true meaning in life is derived from the supernatural, which can also be termed God's purpose theory (Nozick 1981[6]; Craig 1994[7]; Poettcker, 2015[8]). Naturalism, instead, is the theory that meaningfulness is subjectively derived from the individual or objectively derived from a series of

[5] Attoe Aribia (ed). *African Perspectives to the Question of Life's Meaning*, (London and New York: Routledge, Taylor & Francis Group, 2024).

[6] Robert Nozick. *Philosophical Explanations*. (Cambridge, Massachusetts: Harvard University Press,1981)

[7] William Craig. "The Absurdity of Life without God." 1994. https://rintintin.colorado.edu/~vancecd/phil3600/Craig.pdf

[8] Jason Poettcker, Jason. "Defending the Purpose Theory of Meaning in Life". *Journal of Philosophy of Life*, Vol.5, No.3 (October 2015):180–207.

actions that are recognized as meaningful by communities of individuals.[9] Nihilism, on the other hand, is a theory of meaninglessness, stating that living is a futility and achieving meaningfulness is impossible (Schopenhauer 1981[10]; Camus 1955[11]; Martin 1993[12]). To some extent, both supernaturalism and naturalism align with some aspects of African theories of meaningfulness, but not nihilism, as we shall shortly show.

African philosophy of meaning, on the other hand, has been summarized by Attoe (2024) as having basically four theories: God's purpose theory; vital force theory; communal normative functional theory; and consolationist theory.[13] According to Attoe, God's purpose theory means fulfilling one's destiny by obeying divine law to sustain a relationship with the Supreme Being as the basis of meaningfulness in life. Vital force theory posits that meaningfulness is characterized by positive engagement with others and the rituals that enhance one's level of vital force. In partial support of this theory, Metz believes that vital force theory promotes life's force and the force of others, though he thinks this might be too narrow for a good theory.[14] Ada Agada (2020) defends African vital force theory on the meaning of life. Thirdly, the communal normative function theory posits that meaningfulness in life is grounded in communal harmony, which fosters positive relationships with others within the community. Motsamai Molefe supports this theory by stating that personhood and a meaningful life in African philosophy are an earned and normative pursuit, or an achievement of moral excellence and the expression of moral virtues.[15] Monday Igbafen (2017) supports this theory, concluding that life is meaningful within the context of fostering communal relationships with others in society.[16] The communal theory is also supported by Metz, who states that meaningfulness in the communal view involves promoting a communal relationship with other persons. Lastly, consolationist theory maintains that a meaningful life involves the yearning for the pursuit of perfection. Ada Agada (2015) also clearly supports this theory, stating that the

[9] Attoe Aribia (ed). *African Perspectives to the Question of Life's Meaning*, (London and New York: Routledge, Taylor & Francis Group, 2024).

[10] Arthur Schopenhauer. "On the Sufferings of the World," in *The Meaning of Life*, ed. E.D Klemke (Oxford: Oxford University Press, 1981, 45).

[11] Albert Camus. *The Myth of Sisyphus and Other Essays*. (New York: Vintage Books. 1955).

[12] Raymond Martin. "A Fast Car and A Good Woman." In: *The Experience of Philosophy*, edited by D. Kolak and R. Martin, 556–562. (Belmont: Wadsworth, 1993).

[13] Attoe Aribia (ed). *African Perspectives to the Question of Life's Meaning*, (London and New York: Routledge, Taylor & Francis Group, 2024).

[14] Thaddeus Metz. *Meaning in Life*. (Oxford: Oxford University Press, 2013). doi: 10.1093/acprof:oso/9780199599318.001.0001

[15] Motsamai Molefe. *Ethical Reflection on Population Challenges*, Inclusive Society Institute, 2022.

[16] Monday Igbafen. "Human Life and the Question of Meaning in African Existentialism." In: *Themes, Issues and Problems in African Philosophy*, edited by Isaac Ukpokolo, 237–254. (Cham: Palgrave Macmillan, 2017). doi:10.1007/978-3-319-40,796-8_16

consolationist perspective is where meaning is found in the yearning for the pursuit of perfection.[17].

In my view, to understand the theory of meaningfulness in life, we must consider two realities: that humans are individuals in their own right, but that they are also ontologically related to others, nature, and the spiritual world. Due to these individual and ontological relationships, humans have needs at three distinct levels: the physical, the social, and the spiritual. At the level of the body, humans need the material things of this world, food, shelter, and clothing, which are economic needs. However, at the societal level, humans require the society within which they are born, thrive, and find love and acceptance. At the level of the soul, which is the immaterial dimension of man, humans need God.

The weakness in Western philosophy is that, even when it discusses supernatural theories, it tends to isolate them from other equally important dimensions of human life. When they discuss naturalism, for example, they think it can stand alone. As Thaddeus Metz mentioned earlier, to develop a comprehensive theory of the meaningfulness of human life, we need a combination of theories.[18] Our view is that African philosophy seeks to unify these diverse theories into a harmonious system. The weakness of some African theories, such as those proposed by Attoe, is that they are taken as a stand-alone. The African theory of meaningfulness of human life maintains that no true meaning can be achieved in human life if any of these proposals are taken in isolation. This is due to African metaphysics, which posits that humans are individuals ontologically related to others, nature, and the spiritual world. In order to find meaning in life as individuals, humans need the material goods of this world; they need other human beings, nature, and God. The fulfillment of the material needs of humans will need to be supplemented by social ties, where both wealth and poverty are shared. The society supports the individual's struggle to find and process the natural resources of this world. However, this can only be achieved when they respect the environment in which they live. Humans and the natural environment are supplementary to each other, because the carbon dioxide humans breathe out is consumed by the plants, and the oxygen the plants exhale is, in turn, consumed by humans and other animals.

Western naturalism does not want to hear about God and his influence on human life. Some even refer to it as metaphysics: realities that cannot be verified by human science. Therefore, the African theory of meaningfulness in life can only be a harmonious blend of the elements mentioned above. What is important is moderation in understanding these various elements; the exaggeration of faith that everything should be left to God alone is wrong, because humans must work closely with one another, applying the vital forces within them to derive what they need. Equally, denying God's existence is depriving creation of meaning in life, since it leads to

[17] Ada Agada (2020). *Existence and Consolation: Reinventing Ontology, Gnosis and Values in African Philosophy*. (St Paul: Paragon House, 2015).

[18] Thaddeus Metz, *Meaning in Life*, 2013.

reckless exploitation of natural resources, which is counterproductive to the building of meaningfulness in human life.

4.3 An African Perspective on the Meaning of Life

This section deals with the meaning of human life. This chapter questions whether the meaning of life is determined by who is born and how many are born in this world, or whether these factors are inconsequential to a fundamental understanding of the meaning of life.

Susan Wolf defines the meaning of life as "fitting fulfilment."[19] In her words, "meaning in life arises when subjective attraction meets objective attractiveness, and one is able to do something about it or with it."[20] Similarly, Stephen Macedo, in the introduction to Wolf's book, observes that the three crucial elements of this fitting fulfilment are: "subjective attraction, objective worthiness, and active productive engagement."[21] Macedo explains that, according to Wolf, human beings long for fulfilment, and we admire people whose lives are lovingly and productively engaged with projects that are worthy of engagement. But exactly how we should understand the different elements of meaningful lives remains a question to probe. Macedo, however, concludes by saying the view Wolf develops, "supports what ordinary people often say about the importance of living meaningful lives, and so, Wolf 's account helps vindicate much human striving."[22]

In this sense, Wolf seems to underscore the fact that true meaning in life is realized when human subjective attraction and striving meet objective worthiness through engaged, active productivity. In this way, Wolf is saying, meaningful life is the achievement of objective worthiness through an engaged subjective striving. In this understanding, I can see three things: **an end** toward which an individual is striving, which is the objective values embedded in that end; **the striving agent,** who is subjectively attracted toward that end; and **the means used,** which is the active involvement in achieving the perceived end.

Though this sounds perfect, it is unclear what makes such an end an objective value. That humans get subjectively attracted toward some end is clear, but what makes such an end objectively worthy is not really clear. And this is because what gives meaning to life is difficult for most people to determine. Is it personal fulfilment of one's desires, or the fulfilment of societal desires? Western philosophy tends to value the former, while African philosophy tends to value the latter. This is because several thinkers from the Western philosophical tradition think that what is

---

[19] Susan Wolf. *Meaning in Life and Why it Matters*, (Princeton, New Jersey: Princeton University Press, 2010, 25).

[20] Wolf, *Meaning in Life and Why it Matters*, 2010, 26.

[21] Stephen Macedo. "Introduction", in S. Wolf. *Meaning in Life and Why it Matters*, pp. xi-xvi), (Princeton, New Jersey: Princeton University Press, 2010, xii).

[22] Stephen Macedo. "Introduction", in S. Wolf. *Meaning in Life and Why it Matters*, 2010, xii.

important is an individual's striving for excellence and self-fulfilment. By contrast, a number of African philosophers think it is the value of society and its fulfilment that characterizes true fulfilment in human life, which is the objective of human striving.

My view, however, is that we need both personal fulfilment and communal fulfilment. To some extent, if genuine personal desires are fulfilled, there is a likelihood that society's fulfilment will also be fulfilled. The problem comes if there is a clash between the two. Ayn Rand (1964) would simply say that, whenever such conflict occurs, we must choose the personal interest.[23] Plato, by contrast, would say that, when such a conflict occurs, the individual's interest has to be sacrificed. An Acholi adage from northern Uganda, one of the African societies with which I am well acquainted, would say both are equally important. This is captured by one of their proverbs, which states that, "*Mak atum pa rwot med ki megi*": "Hold the weapon of the king and add also your own."

The meaning of this adage is that while the king is a symbol of social authority and interest, whose weapon is very important for the defense of society, the servant also has to take care of his own security, in the event of getting into the thick of a battle. In such a scenario, the king will take up his own weapon to defend himself; the servant must not be left vulnerable without a weapon of defense. This implies that both personal interests and social interests are important and are objectively worthy. The objective worthiness Wolf is referring to should balance these two interests in the process of living. So, personal interests are equally important as social interests; objectivity is realized when neither of these two interests is unrealistically or unjustly compromised.

In fact, the three issues identified by Wolf remain important, but her understanding and mine are not the same. First, I agree with her that personal drives and desires are important, but I would add that personal desires must be balanced with the needs and desires of society in which an individual lives. Secondly, the end must be some objective worthiness, but my understanding of this objective worthiness is a combination of personal and social interests. Thirdly, while for Wolf the means of realizing this end is through personal engagement, I think we still need a combination of personal engagement with social support. However heavily engaged we are in the fulfilment of our personal interests, at the same time, we need the support of society within which we live. This would cater for the vulnerable members of our society who may be hampered from becoming personally engaged.

In general, Western society has better standards of living and a higher quality of human welfare, but this is not because of individual strengths and personal engagements alone. There are very good social security systems in the West, supplementing the efforts of individuals in a number of developed countries. Without such support, most developed countries' standards of living would automatically drop, and the disparity in the standards of living would reach possibly unsustainable levels. It was within this social support system that most traditional African societies

[23] Rand Ayn. *The virtue of selfishness: A new concept of egoism*. (New York: Penguin Books, 1964).

met their security and welfare needs, which gave them meaning in life. A number of modern African societies have now lost this precious niche in life, and this has left a number of people within Africa grappling with meaninglessness in life. Nyerere (1968) describes it as personalized wealth and privatized poverty; since when both wealth and poverty are shared, there would be less suffering in the world. In sharing, wealth becomes more fulfilling and poverty less burdensome.

In addition to social support, African traditional societies never lost their connection with their environments, including nature and the spiritual world. This, too, gave them a very strong basis for a meaningful life. Today, modern science thinks it can manipulate and replace natural products within our environment with artificial ones; they are getting this wrong.

4.4 Population Discourse and Meaning of Life

Our argument has been that population ethics investigates the various ethical issues that arise when we consider how our actions affect both "who" is born and "how many" people are born.[24] In this section, we are analysing the relations between population ethics and the meaning of life.

4.4.1 About Who Is Born

Western population discourse is very much concerned about who is born. This is because who is born determines the quality of life that he or she might live in the future. When one is born into a poor family or community, the likelihood is that that person will inherit the poverty of the society in which they are born and end up living a life of poverty and a low quality of life. The implication is that it would be better to reduce such possibilities so that better-off persons, whose quality of life would be higher, are the ones born, so that poverty in the world is reduced and the world is made, qualitatively, a better place to live in.

This argument implies that population growth in the poor countries and continents of the world has to be addressed, since it is they who are likely to increase burdens on the world that would have been a better place to live in. In such an argument Africa becomes, undoubtedly, the continent in question. Statistics generated by population experts show that the trends of population increase in Africa will cause the world a big price to pay. No wonder Ehrlich's "The Population Bomb" became the epitaph of such statistical data.

This also implies that disabled persons are a burden to society, and it may be better that such individuals be eliminated before they become a burden too big to

[24] Hilary Greaves. "Population axiology". *Philosophy Compass*, 12 (11), 2017, e12442.

handle. Some of the measures to reduce such risks, among others, include: Mercy Killing for those terminally ill, abortions, and infanticide seem to be some of the measures employed. With regard to Mercy Killing, though this is a practice existing in the world, not out of documented policy, the question they ask is why one should be allowed to suffer so much from a disease that is already known to be incurable, instead of helping such a person to end their life in a peaceful and less painful way. Johann Zwirner et al. (2025: 392) believe that mercy killing, which is sometimes referred to as euthanasia, still exists in "eight jurisdictions worldwide." At the back of such a question is the argument that meaning in life can only be achieved, as Wolf puts it, when an individual is engaged, and a terminally ill person cannot be engaged. The second implication is that illness and pain are contrary to human wellbeing, so there is no way an individual who is terminally ill can be happy, because he or she lacks both meaning in life and objective worthiness.

4.4.2 About how Many People Are Born

This is an even bigger question. How many people can our continent contain in order that everybody is able to live a productive and meaningful life without poverty and with high standards of living? The answer to this question is always the same: a smaller number is needed in the world. An associated question would be: What should we do to reduce the world population? The answers to this are varied; some would say target human fertility, including the termination of unborn children, or even infanticide, develop policies for single-child families as was the case in China for a number of years, or at worse, replace humans with robots, artificial persons (AP) or artificial intelligence (AI). Some would say, produce more food. But how much more food can the world produce for a continuously growing human population? The prospects are genetically engineered and technically modified foods. All these methods are being employed, yet the degree of their success can never be established with any certitude.

Even if evidence has been given by authors such as David Lam (2023) that Ehrlich's "Population Bomb" has been proved false, they still believe it. In the view of Ehrlich, the world population of 3 billion by then (the 1960s) would burst and create a bomb-like impact on the world by causing mass starvation. But 50 years later, the world population increased by 4 billion, and yet the bomb has not exploded, as has been mentioned earlier.

My problem is still with population discourse and the meaning of life. The meaning of life is not just created by abundance as assumed by utilitarianism, which is the maximization of wealth and the minimization of suffering and poverty. In my view, the greatest poverty is not the lack of material goods, but a combination of many things, among which are: lack of material goods, lack of meaningful human relationships, and the lack of inner peace, which would be created by good relations with the environment and the invisible world.

I concur with the Christian Bible, where, in the Acts of the Apostles, it is written, “None of their members was ever in want” (Acts 4:34). Here, not to be “in want” means one lacks nothing. The reason they give is that “all those who owned land or houses would sell them, and bring the money from them, to present to the apostles; it was then distributed to any members who might be in need” (Acts 4:34–35). In addition, “The whole group of believers was united, heart and soul” (Acts 4:32). They loved one another, and they always met in prayer. So, for one to lack nothing in this world, we need three things: the material wealth of this world; each other; and the knowledge that we are in need of God. Whenever any one of these three is missing, we lack true meaning in life.

4.5 African Metaphysics as a Basis for a Meaningful Life

In this chapter, I have more belief in African metaphysics as a basis for making any meaningful value judgments on the life of any human population in as far as who is born and how many people are to be born is concerned. The first concern is in deriving policies regulating how many people are to be born. The value of human life, as mentioned above, is defined by the purpose for which life is given to humans. It does not matter whether this purpose is given by God or embedded in human nature, or even whether individuals attempt to design their own purpose for living. What is important for me here is the purpose itself.

Borrowing from one of the wise sayings of the Acholi people of Northern Uganda, we can understand the purpose of life as buried within humanity itself. The purpose of life from this adage, which says *Obwol top ka ngwinye*—meaning “Where a mushroom sprouts is where it rots.” Meaning, deeply buried within nature itself is the principle of *causality* and *finality*. So, buried within nature is the ability to come to be, but at the same time to cease to be. Similarly, humans, like any other natural beings, have within themselves the principles of *causality* and *finality*. Any attempt to manipulate nature gravely undermines and interferes with these basic principles of life, to be and to cease to be.

In the view of this chapter, the purpose of life is the harmonious existence of an individual with other individuals within society, harmony with nature and the environment, and harmony with the spiritual world. For this harmony to prevail, humans need to be allowed to exist in accordance with their nature. This is because humans need the material wealth of this world, they need one another, and they need God. But how can one achieve this end? Our view is that it takes more than the engaged involvement of an individual, since an individual’s effort is to be supplemented and supported by the society within which those individuals live. The end must take cognizance of both personal interests and social interests.

The Global Footprint Network (GFN) holds that “A footprint in excess of 1 means demand for ecological services exceeds their supply.” By GFN’s reckoning, “the world’s average living standard at the level reached some ten years ago (roughly 12,000 international dollars) would have required 1.5 Earths. It is now evident that the enormous gains humanity has enjoyed in the past 65 years have been at the expense of a diminished biosphere we are leaving behind for our descendants."

Ehrlich and Holdren (1971) proposed a formula to understand this reality, which stated that: I=PAT. to draw attention to the significance of the biosphere's carrying capacity for population ethics. The authors traced the ***i***mpact of human activities on the Earth system to ***p***opulation, ***a***ffluence (read, per capita consumption of goods and services), and the character of ***t***echnology in use (including institutions and social capital). Because our impact on the biosphere is proportional to the demands we make of it, and because those demands increase with our level of economic activity, we can assume our impact on the biosphere increases with economic activity.[25]

The implication is that any rise in population increases the per capita consumption of goods and services (affluence), including technology in use, which impacts the Earth's biosphere. This means that population increase has a global footprint, which is usually negative. In this chapter, this means not negating or disputing this scientific fact, but the interpretation of this scientific data and the method used to address this scientific information. The African perspective is not against this reality, but rather against the interpretation of this fact. In this instance, the argument is that it is the poor populations who are basically responsible for the unregulated increases of the world population. This is clearly demonstrated by Table 4.1,[26] which is presented in the Appendix to this chapter.

The table indicates that the fertility rates of rich countries of the world are around 1.7, while those of poor countries are 4.9. Secondly, the GDP per capita of rich countries is 41,000 international dollars, while that of poor countries is 1570 international dollars.

This means, using Ehrlich and Holdren's (1971) formula, A stands for Affluence, and T stands for technology, that there is a problem with the poor nations of the world.[27] Here, the implication is that a greater impact on the Earth's biosphere comes from the rich countries of the world, since they are more affluent than the poor countries of the world. By contrast, population increase is caused by poor countries because of their high fertility rates. Thirdly, technology, as one of the causes of a bad impact on the Earth's biosphere, comes from rich countries, while the poor countries of the world make a relatively small contribution to it. In the final analysis, the policies in the world are skewed toward a reduction in population, since it is believed that it is the increase in population that creates substantial damage to the Earth's biosphere. This is a mismatch which has been discussed by this chapter. Less emphasis is placed on affluence and technology as having damaging impacts on the Earth's biosphere.

[25] Aisha Dasgupta and Partha Dasgupta. "Population Overshoot". In K. Bykvist and T. Campbell, eds., *Oxford Handbook of Population Ethics*, (Oxford: Oxford University Press, 2018, 3).

[26] Source: World Bank. (2016). UNPD (2015), Freedom House (2017. For details see Appendix 1.

[27] Paul R. Ehrlich, & John P. Holdren. *Impact of Population Growth in Science*, New series, Vol 171, No. 3977, 1971, 1212–1217.

Another implication is that within human societies, there are those who cannot become personally involved to achieve their interests because of various limiting factors caused by the environment, disabilities, and status of an individual, such as being a child, for instance. Thus, the purpose of life is the fulfilment of the self in the context of society, by attaining material, social, and spiritual interests. It requires not just self-fulfilment, but also the fulfilment of the interests of society. It demands that the human population should be regulated, but not manipulated by means that undermine the integrity of the human person, or that even find them replaced by APs or AI.

Finally, the implication is also that the quality of human life is not determined by who is born and how many people are born. It is related more to how the individual and society handle those that are born, and meaningfully design means that are ethically acceptable to regulate how many people are to be born. The means used and the end of what is designed to regulate the human population must also be ethically acceptable. Other ethical theories, such as utilitarianism, are insufficient to provide this guidance in terms of both the means and the end. African metaphysics, because of its insistence on harmonious living with others, the environment, and the spiritual world, provides a better understanding of the meaning of life in this world.

4.6 Conclusion

The question of the meaning of life is answered when humans realize that they are more than atomic individuals; they are also social beings. As individuals, they have personal interests and desires, and potential to strive toward self-fulfilment. As social beings, humans cooperate with one another, supporting and sustaining those less fortunate, and also realizing their interests and self-fulfilment. To find meaning in life, humans need the individual, society, and the spiritual world. Equally important is the fact that there can be no meaning in life without the satisfaction of our material needs, in addition to our social and spiritual needs.

Secondly, attempts to regulate the human population are acceptable, but the means used must be ethically justifiable. This is because it is not sufficient that the end should justify the means; rather, both the means and the end must be ethically justifiable. To achieve a good end, we must equally employ good means. African metaphysics supports this ethical theory. In trying to regulate the human population, we must not deny existence to humans by replacing them with APs and robots or AI; neither should we destroy the environment within which we live, since this is the womb of human existence. Lastly, it is not by destroying others that our well-being and quality of life will be improved. Technically, our welfare is not dialectically opposed to the existence of others as actual or potential competitors. Sharing, cooperation, working together, and personal self-engagement in the achievements of our goals must be supplemented with the support of society and the spiritual world.

Appendix 1

Table 4.1 Social statistics from rich and poor regions (Year 2014–15)

	Rich	poor
Population (millions)	1420	620
GDP per capita (international dollars)	41,000	1570
Total fertility rate[a]	1.7	4.9
Under-5 mortality rate (per 1000)	7	76
Life expectancy at birth (years)	80	60
Youth literacy	100	68
Civil liberties	High	Low
Political liberties	High	Low
Government corruption	Low	High

Source: World Bank (2016), UNPD (2015), Freedom House (2017)
[a] Total fertility rate (TFR) is the number of births that a woman expects to have during her reproductive years. The number 2.1 is usually taken to be the TFR that, over the long run would lead to a stable population

References

Ada Agada (2020). The African vital force theory of meaning in life, South, African Journal of Philosophy, 39:2, 100–112, Retrieved from: https://doi.org/10.1080/02580136.2020.1770416

Agada, Ada. 2015. *Existence and consolation: Reinventing ontology, gnosis and values in African philosophy*. St. Paul: Paragon House.

Dasgupta, Aisha, and Partha Dasgupta. 2018. Population overshoot. In *The oxford handbook of population ethics*, ed. K. Bykvist and T. Campbell. Oxford: Oxford University Press.

Attoe, Aribiah, ed. 2024. *African perspectives to the question of life's meaning*. London and New York: Routledge, Taylor & Francis Group.

Ayn, Rand. 1964. *The virtue of selfishness: A new concept of egoism*. New York: Penguin Books.

Camus, Albert. 1955. *The myth of Sisyphus and other essays*. New York: Vintage Books.

Craig, William. 1994. *The absurdity of life without god*. https://rintintin.colorado.edu/~vancecd/phil3600/Craig.pdf

Ehrlich, R. Paul, and John P. Holdren. 1971. Impact of population growth. *Science* 171 (3977): 1212–1217.

Ehrlich, R. Paul. 1968. *The population bomb*. 12th ed. New York: Ballantine Books.

Greaves, Hilary. 2017. Population axiology. *Philosophy Compass* 12 (11): e12442.

Igbafen, Monday. 2017. Human life and the question of meaning in African existentialism. In *Themes, issues and problems in African philosophy*, ed. Isaac Ukpokolo, 237–254. Cham: Palgrave Macmillan. https://doi.org/10.1007/978-3-319-40796-8_16.

Lam, David. 2023. Has the world survived the population bomb? A ten-year update. *Population and Environment* 45 (10): https://doi.org/10.1007/s11111-023-00422-7.

Macedo, Stephen. 2010. Introduction. In *Meaning in life and why it matters, by Susan Wolf, xi–xvi*. Princeton, NJ: Princeton University Press.

Martin, Raymond. 1993. A fast car and a good woman. In *The experience of philosophy*, ed. Daniel Kolak and Raymond Martin, 556–562. Belmont, CA: Wadsworth.

Metz, Thaddeus. 2013. *Meaning in life*. Oxford: Oxford University Press. https://doi.org/10.1093/acprof:oso/9780199599318.001.0001.

Molefe, Motsamai. 2022. *Ethical reflection on population challenges*. Inclusive Society Institute.

Nozick, Robert. 1981. *Philosophical explanations*. Cambridge, MA: Harvard University Press.

Nyerere, J. K. (1968). Ujamaa: Essays on Socialism, Dar es Salaam: Oxford University Press

Jason, Poettcker. 2015. Defending the purpose theory of meaning in life. *Journal of Philosophy of Life* 5 (3): 180–207.

Schopenhauer, Arthur. 1981. *On the sufferings of the world.* in The Meaning of Life, ed. E.D Klemke (Oxford: Oxford University Press, 1981).

The World Bank. 2016. *The world bank report 2016*. New York: World Bank.

Wolf, Susan. 2010. *Meaning in life and why it matters*. Princeton, NJ: Princeton University Press.

Zwirner J, Iwersen-Bergmann S, Püschel K, Ondruschka B. (2025), Mercy killing of a 72-year-old woman through heroin intoxication. *Journal of Forensic Sciences* 70:392–397. https://doi.org/10.1111/1556-4029.15637

Chapter 5
Tension Between Procreation as a Duty and Individual Right in African Population Ethics

Nomatter Sande

Abstract This chapter explores procreation as an ethical and cultural construct within African communities, where it transcends individual reproductive choice and becomes a communal duty rooted in spiritual beliefs, lineage obligations, and socio-economic practices. Drawing on ethnographic examples from various African cultures, such as the Shona, Yoruba, Dagomba, Maasai, and Akan, the chapter illustrates how procreation serves not only to sustain family and lineage but also to uphold social order and ancestral continuity. It contrasts these indigenous perspectives with emerging reproductive rights discourses across the continent, highlighting the tensions between tradition and modernity. Through the conceptual lens of moral demography, the study reveals that demographic behavior in Africa cannot be disentangled from local moral economies, historical influences such as colonization and Christianity, and the evolving roles of institutions like the church and feminist movements. The chapter concludes that understanding African population ethics necessitates moving beyond biomedical and Western frameworks to incorporate indigenous moral reasoning and socio-cultural dynamics.

Keywords Population ethics · Procreation · The right to procreate · African indigenous people's rights · African human rights charter

5.1 Introduction

In modern African communities, the conflict between reproduction as a social obligation and as an individual right presents a fundamental ethical conundrum. The United Nations projects that Africa's population will double to 2.5 billion people,

N. Sande (✉)
University of South Africa, Institute of Theology and Religion (RIIT), Pretoria, Gauteng, South Africa

B. Okyere-Manu, P. Mwipikeni (eds.), *An African Approach to Population Ethics*, Advancing Global Bioethics 23,
https://doi.org/10.1007/978-3-032-24464-2_5

representing an unparalleled growth in human history.[1] Yet, behind these statistics lies a complex tapestry woven from culture, historical legacies, and even ethical dilemmas. For instance, in Zimbabwe, childless individuals are buried with a rat to appease ancestors, thereby symbolizing the weight of procreation.[2]

Although this practice does emphasize the spiritual weight of childlessness, it also raises important issues regarding the psychological burden faced by those dealing with infertility in contemporary settings. How can such behaviors coexist with new discourses on mental health and physical autonomy? This conflict between ancestral needs and modern human rights systems is particularly evident when forced childlessness shifts from a personal situation to a societal stigma.

There needs to be a greater understanding of the reasons behind this growth and the foundations of population ethics, the study of which explores the ethical issues arising from human actions concerning their impact on future generations and the number of people born.[3] This field encompasses vital aspects such as wellbeing, welfare, and diverse perceptions of happiness within a population. Population ethics can be broadly categorized into two axes: first, axiology, which addresses the number, welfare, and identity of existing people; and second, normative aspects, which examine the moral implications of creating individuals with varying qualities of life. This chapter concentrates primarily on normative ethics. It is worth noting that most discussions and academic definitions of population ethics originate from Western perspectives. While these viewpoints are valuable, it is essential to recognise that different cultures possess unique values, histories, and beliefs that shape their understanding. Considering this status quo as a starting point to comprehend population ethics, it is insufficient to assume that Western categories adequately address African population dynamics. Although the increasingly globalized world continues to enhance Western notions of population ethics, intrinsic pockets of resistance persist, manifesting in diverse customs, beliefs, and hybrids of the two dispositions. This chapter focuses on the right to procreation.

When reviewing the term 'procreation', it should be observed that it bears a close affinity to 'reproduction.' Christian theology popularizes procreation, referring to the begetting of God the Son by God the Father.[4] Medieval Christianity adopted the norms and rules concerning marriage, sex, and procreation.[5] According to Strathern, "depicted sexual intercourse as a core symbol: the diffuse enduring solidarity of

---

[1] United Nations, World Population Prospects 2019: Highlights (New York: United Nations, 2019).

[2] Nomatter Sande and Tawanda Takawira-Mtwaya, "Assisted Reproductive Technologies, Infertility and Women in Zimbabwe," in Religion, Women's Health Rights, and Sustainable Development in Zimbabwe, ed. Sophia Chirongoma, Molly Manyonganise, and Ezra Chitando (Cham: Springer, 2022), 45–62.

[3] Gustaf Arrhenius et al., "Introduction," in The Oxford Handbook of Population Ethics (Oxford: Oxford University Press, 2022), 1–12.

[4] Nick Hopwood, Rebecca Flemming, and Lauren Kassell, eds., Reproduction: Antiquity to the Present Day (Cambridge: Cambridge University Press, 2018).

[5] Michel Foucault, Histoire de la sexualité 4: Les aveux de la chair, ed. Frédéric Gros (New York: Gallimard, 2018).

close family relations was attributed to sharing substance through the act of procreation. Procreation was a natural fact of life."[6] The notion that procreation is portrayed as a natural reality introduces the word 'reproduction' as a biological process.[7] In broader terms, Strathern summarizes the difference, stating that "'reproduction' has to do with replicating an original, not identical but like the original, with producing offspring of one's kind, whereas 'procreation' has to do with the 'act of begetting', the 'effectiveness of the capacity' to produce offspring."[8] With this understanding of procreation, a prominent question in this chapter is to what extent Africans have the right to procreation, what liberties they have to choose when, whether, and how many children to have.

This spans population dynamics Sayer[9]; Johnson-Hanks[10]; to Thompson's[11] moral economy, which analyses how normative institutions control economic activities. By means of this framework, the essay contends that knowledge of African population ethics calls for a transcending of Western biological and rights-based paradigms to engage indigenous moral reasoning and hybridized socio-cultural reality.

As such, aspects of moral demography help to understand African populations. In this chapter, moral demography, that is, an analytical framework that examines how moral standards (e.g., honor, ancestral obligation) interact with socio-historical conditions to determine demographic behaviour, is applied. Emphasizing how communal ethics guide reproductive decisions within changing conditions, it expands Thompson's[12] idea of moral economy, which studies how normative institutions control economic actions, into population dynamics (Sayer[13]; Johnson-Hanks[14]).

This chapter addresses a crucial gap: the unresolved conflict between the rising discourse of individual reproductive rights shaped by modernity, feminism, and global human rights frameworks, and the non-negotiable obligation rooted in African traditional ethics, governed by lineage, ancestral continuity, and communal

---

[6] Marilyn Strathern, After Nature: English Kinship in the Late Twentieth Century (Cambridge: Cambridge University Press, 1992a), 43.

[7] Strathern, After Nature.

[8] Marilyn Strathern, "Postscript: A Relational View," in Technologies of Procreation: Kinship in the Age of Assisted Conception, second ed., ed. Jeanette Edwards et al. (London: Routledge, 1993), 208.

[9] Andrew. Sayer 2000. "Moral Economy and Political Economy." Studies in Political Economy 61 (1): 79–104.

[10] Jennifer Johnson-Hanks, "Uncertainty and the Second Space: Modern Birth Timing and the Dilemma of Education," European Journal of Population 20 (2004): 351–373.

[11] Edward P. Thompson, 1971. "The Moral Economy of the English Crowd in the Eighteenth Century." Past & Present 50 (1): 76–136.

[12] Edward P. Thompson, 1971. "The Moral Economy of the English Crowd in the Eighteenth Century." Past & Present 50 (1): 76–136.

[13] Sayer, Andrew. 2000. "Moral Economy and Political Economy." Studies in Political Economy 61 (1): 79–104.

[14] Jennifer Johnson-Hanks, "Uncertainty and the Second Space: Modern Birth Timing and the Dilemma of Education," European Journal of Population 20 (2004): 351–373.

welfare. To unpack this tension, the chapter first establishes procreation as a socio-spiritual imperative across African societies, drawing on ethnography from cultures such as the Shona, Yoruba, and Maasai. It then examines how modernist forces, including urbanization, economic pressures, reproductive technologies, and feminist activism, are reshaping traditional ethics, creating friction with communal norms. As a context-sensitive lens, it finally suggests moral demography, defined here as an analytical framework that analyses how moral standards, such as honor and ancestral responsibility, interact with socio-historical contexts, including colonialism and institutional change, to affect demographic behavior.

5.2 Understanding the Need for the Right to Procreation

It is essential to justify how and why the right to procreation is critical. In African population ethics the right to procreation is a significant issue that merits scholarly attention because, with ongoing concerns about overpopulation, resource allocation, and sustainable development, this has become an increasingly topical debate.[15] These concerns pose ethical dilemmas surrounding the balance between individual reproductive rights and broader societal needs.[16] Procreation is deeply intertwined with lineages, social identity, and collective values; to a greater extent, communal values contrast with contemporary individualistic reproductive rights.

The right to procreation raises legal and human rights issues.[17] The effectiveness of population controls established and imposed during colonial times must be reassessed. Laws should evolve in response to changes in societal norms, technological advancements in reproductive health, and international human rights discourse. Another aspect concerns gender and social equity. Procreation rights are closely tied to issues of gender equity, particularly in societies where cultural, religious, or legal constraints may limit women's reproductive choices.[18] Scholarly attention is needed to explore how these dynamics manifest in African contexts and to advocate for rights that support individual autonomy and social equity.[19]

The right to procreation must be understood in relation to technology and medical advances. The advent of new reproductive technologies, such as IVF and genetic screening, poses new ethical challenges, particularly in African settings, where

[15] Akira Yamada, ed., Population and Development in Africa: The Role of Indigenous Knowledge (Tokyo: United Nations University Press, n.d.).

[16] Jason L. Finkle and C. Alison McIntosh, "The New Politics of Population: Conflict and Consensus in Family Planning," Population and Development Review 20 (1994): 3–38.

[17] African Union, African Charter on Human and Peoples' Rights (1986), OAU Doc. CAB/LEG/67/3 rev. 5.

[18] Emmanuel Chukwudi Eze, ed., African Philosophy: An Anthology (Oxford: Blackwell, 1997).

[19] Joanna N. Erdman, "Reproductive Rights in Africa: Strategies and Approaches," International Journal of Gynecology & Obstetrics 119 (2011): 198–202.

there may be variations in access, cultural acceptance, and regulatory frameworks.[20] These technologies raise questions about the right to procreate in contexts where traditional practices are profoundly valued.[21] Additionally, there is the issue of unequal access to reproductive healthcare, which has an impact on the realization of the right to procreate. Scholarly attention is needed to address how these inequalities can be alleviated in African populations.

The right to procreation encompasses both philosophical and theological considerations. African ethical philosophies, such as Ubuntu, emphasize community and mutual care, which may provide unique perspectives on the right to procreation.[22] These philosophies offer alternative ways of understanding reproductive rights that differ from Western individualistic frameworks. In many African contexts, religious beliefs significantly shape attitudes toward procreation. Theological perspectives on procreation, which vary widely across African religions, must be examined to understand their impact on ethical discussions and policymaking.

5.3 Trajectories of the African Population

An understanding of the population dynamics of Africa is crucial for effective policy planning, as population growth has a significant impact on livelihoods, global health, and sustainable development.[23] Reconstructing the history of the African population is complex due to the impacts of slavery and colonization. European historical data primarily relies on church records.[24] Walter summarized the complexity as "it is known about even the relatively recent demographic history of Africa because of the lack of data.[25] Elsewhere, historical demographic trends have been reconstructed by applying family reconstitution to church records." Three trajectories are prominent, commencing with the nineteenth century, continuing into the twentieth century, and extending into the current twenty-first century.

The demographic trajectory of the nineteenth century is characterized by an increase in slavery, epidemic diseases, and epizootic diseases.[26] Although basic research projects 12.5–15.4 million enslaved Africans shipped over transatlantic

[20] Kwasi Wiredu, "The African Concept of Personhood," in The Stanford Encyclopedia of Philosophy, ed. Edward N. Zalta (2020).

[21] Sande and Takawira-Mtwaya, "Assisted Reproductive Technologies."

[22] Thaddeus. Metz 2014. "Ubuntu and the Debate on African Philosophy." African Studies Review 57 (2): 21–39.

[23] John Cleland and Kazuyo Machiyama, "The Challenges Posed by Demographic Change in Sub-Saharan Africa: A Concise Overview," Population and Development Review 43 (2017): 264–286.

[24] Cheikh Mbacké, The Persistence of High Fertility in Sub-Saharan Africa: A Comment (2017).

[25] Sarah Walter, "Counting Souls: Towards an Historical Demography of Africa," Demographic Research 34 (2016): 63–108.

[26] Helge Kjekshus, Ecology Control and Economic Development in East African History: The Case of Tanganyika 1850–1950, 2nd ed. (London: James Currey, 1996).

routes,[27] current syntheses, for example, Nunn, confirm this figure and its long-term demographic effects.[28] The impact of the slave trade remains debated due to the lack of precise starting population figures. Factors such as mortality, reproductive potential, age, and sex of slave populations, along with matrilineality and polygamous practices, all play crucial roles.[29] These historical disruptions, particularly slavery, epidemics, and colonialism, reshape the African kinship system. To a greater extent, they amplify the cultural emphasis on procreation as a means of resilience. For instance, colonial tax policies encourage people to have larger families. Doyle found that large families were an investment in the demand for labor for cash crops, recasting children as economic assets.[30] This historical context underscores why, in the present day, procreation is inevitably seen as a communal duty rather than an individual choice in most African communities.

Challenges in the nineteenth century also included famines, influenza, smallpox, and sleeping sickness.[31] Historical gaps exist where records should have been prioritized. The population growth of the twentieth century was characterized by increased fertility during the 1920s and 1930s, driven by changes in production and reproductive relations. Colonial demands for taxes resulted in children working in family cash crops, early marriages, reduced polygyny, and increased education for women, which challenged traditional birth spacing practices.[32]

Coming up to the present, it is projected that Africa's population will have grown tenfold from 1960 to 2050, with sub-Saharan Africa accounting for over half of this increase.[33] The estimate stands at 2.5 billion people. However, a stagnating fertility transition introduces uncertainty regarding future population trajectories.[34] Governments are concentrating on enhancing reproductive health and promoting voluntary family planning to positively influence population growth. Globally, there

---

[27] Joseph E. Inikori, 1992. The Chaining of a Continent: Export Demand for Captives and the History of Africa South of the Sahara, 1450–1870. Mona: Institute of Social and Economic Research.

[28] Nathan, Nunn 2023. The Historical Roots of Contemporary African Development. Cambridge: Harvard University Press.

[29] Paul Lovejoy, Transformations in Slavery: A History of Slavery in Africa, 3rd ed. (Cambridge: Cambridge University Press, 2011).

[30] Shane Doyle, "Demography and Disease," in The Oxford Handbook of Modern African History, ed. John Parker and Richard Reid (Oxford: Oxford University Press, 2013), 38–55.

[31] Helge Kjekshus, Ecology Control and Economic Development in East African History: The Case of Tanganyika 1850–1950, 2nd ed. (London: James Currey, 1996).

[32] Shane Doyle, "Demography and Disease," in The Oxford Handbook of Modern African History, ed. John Parker and Richard Reid (Oxford: Oxford University Press, 2013), 38–55.

[33] United Nations. 2019. World Population Prospects 2019: Highlights. New York: United Nations Department of Economic and Social Affairs.

[34] Bastian Schumacher, 2019. "Stalls in Fertility Transitions in Sub-Saharan Africa: Revisiting the Evidence." Studies in Family Planning 50 (3): 257–258.

is an expectation for women to assume responsibility for high fertility preferences.[35]

5.4 Procreation as a Construct of African Population Ethics

The overarching question in this chapter is: to what extent do Africans have the right to procreation? What liberties do they have to choose when, whether, and how many children to have? Paying attention to the answers to these questions helps find the entry point to understanding the constructions of population ethics in Africa. This approach resonates with the moral demography theoretical framework, allowing selective elements to reveal population demography. For instance, from the moral demography lens, the decision about reproduction is "seen as comprising a dynamic set of values, like 'tradition', that continually inform people's perceptions of and accommodations to their often-changing conditions."[36] Thus, in this chapter, the values embedded in culture serve as the stepping stones that construct the ethics of the African population.

5.4.1 The Right to Procreation and Dilemmas

In Africa, 'the right to procreation' serves as a starting point for understanding the construction of population ethics. The use of the term 'rights' raises a debate because it is not readily available within African indigenous cultures and traditions. Historically, the adoption of elements of contemporary 'rights' was consolidated in the early 1960s. The Human and Peoples' Rights, also known as the African Charter, comprises 68 articles that encompass most aspects of society, including, but not limited to, civil, political, economic, social, and cultural rights. According to Matumbu, "the African Charter on Human and Peoples' Rights presents a human right (thus including all sexes—males, females, and the intersex)."[37] The introduction of human rights should be viewed as a recent development within the African context, albeit one with initial challenges. For instance, alongside the launch of the African Charter for Human Rights, many African states are advocating for the Rights of Indigenous People. The rights of African indigenous peoples are currently

[35] Jade Sasser, 2018. *On Infertile Ground: Population Control and Women's Rights in the Era of Climate Change*. New York: New York University Press.

[36] Thomas, Spear, 1997. Mountain Farmers: Moral Economies of Land and Agricultural Development in Arusha and Meru. Berkeley: University of California Press, 11–12.

[37] Faith, Matumbu, 2023. "Women's Rights: A Precursor for African Development." In Contemporary Development Ethics from an African Perspective, edited by Beatrice Okyere-Manu, Stephen Nkansah Morgan, and Obvious Katsaura, 157–172. Cham: Springer, 158.

in their infancy, as African states express their cultural concerns.[38] The scope of indigenous people's rights[39]; most African states were reluctant to recognise the specific rights of indigenous people.[40] In 2007, the United Nations Declaration of the Rights of Indigenous People (UNDRP) was finally adopted.[41]

Whatever the circumstances, what is critical in the development of human rights in Africa is the notion and reaffirmation that there is indeed something distinct called 'indigenous peoples' human rights.' The rights of African Indigenous Peoples focus on cultural, spiritual, and social interactions in the land of ancestors.[42] These are all absent from the African Charter of Human Rights; this was the main reason for the development of another charter, that on 'African Indigenous People's Rights,' to address these concerns. Against this backdrop, it is prudent to adopt a casual approach when discussing the 'right to procreate' within the African context, refraining from rigidly adhering to the well-developed notions of human rights within contemporary Western society. It is correct to accept that in the African indigenous cultural context, 'the right' to procreate is communal and not tied to the 'secular legal law' or human rights activism.

The legal weight here should be viewed as residing within the communities and their invisible traditions, vices, and values, which individuals tend to subconsciously recognise as the right things to do. This article chapter that the 'indigenous people's cultural rights' are a dominant source for the construction of population ethics in Africa. The 'right to cultural integrity' concentrates on the rights to culture, subsistence, heritage, religion, and livelihood.[43] To be fair, the order of life is cyclic and celebrated through the rites of passage of being born, marriage, procreation, and after death. Among these rites of passage, procreation holds a salient and strong value that drives the pace of procreation.

---

[38] James, Anaya, 2004. Indigenous Peoples in International Law. Oxford: Oxford University Press.

[39] Hildergard, Weber, Dahl, Jens and Wilson, Fiona and Waehle, Espen eds. 1993. Never Drink from the Same Cup: Proceedings of the Conference on Indigenous Peoples in Africa. Copenhagen: IWGIA.

[40] Albert Kwokwo, Barume, 2009. "Responding to the Concerns of the African States." In Making the Declaration Work: The United Nations Declaration on the Rights of Indigenous People, edited by Claire Charters and Rodolfo Stavenhagen, 72–89. Copenhagen: IWGIA.

[41] Albert Kwokwo, Barume, 2009. "Responding to the Concerns of the African States." In Making the Declaration Work: The United Nations Declaration on the Rights of Indigenous People, edited by Claire Charters and Rodolfo Stavenhagen, 72–89. Copenhagen: IWGIA.

[42] James, Anaya, and Robert Williams. 2001. "The Protection of Indigenous Peoples' Rights over Lands and Natural Resources under the Inter-American Human Rights System." *Harvard Human Rights Journal*. Accessed December 22, 2023. https://scholar.law.colorado.edu/faculty-articles/846.

[43] Jérémie, Gilbert, 2009. "Custodians of the Land: Indigenous Peoples, Human Rights and Cultural Integrity." In Cultural Diversity, Heritage and Human Rights, edited by Michelle Langfield, William Logan, and Máiréad Nic Craith, 31–44. London: Routledge.

5.4.2 Gender Roles and Reproductive Burden

Within the procreation mantra, women are at the center of the discourse, particularly concerning their fertility. Africa's demographic data from numerous countries indicate that women constitute the largest segment of the population. Recent statistics confirm that women make up 50.2% of Africa's population, thereby supporting the findings of previous studies.[44] It can be summarized that "the percentage of women in Africa is more than half the population of the African continent."[45] Population growth is anticipated since women's fertility is regarded as a blessing. Culturally, women are viewed as the incubators of population growth.

The institution of marriage is not only desired, but also regarded as an achievement. Women's pregnancy is not optional; rather, it is celebrated as the fulfillment of the primary purpose of marriage. There is no choice concerning the necessity of having a child. It is anticipated that individuals should conceive and bear children after marrying. Every woman and man is expected to marry at some point. Although bearing children is the principal expectation of society, it outlines family life, joy, and children's experiences as secondary reasons for marital life. Accordingly, this position resonates with notions of population ethics that focus on happiness and well-being.[46]

Indirectly, procreation carries a subtle economic benefit, since most parents still expect their children to bear the burden of their aging. In this regard, procreation is a duty framed as a natural function of women. Large families are viewed as a blessing due to the economic empowerment they bring. Traditional African communities regard children as blessings from the supreme being. Ancestors are central to creation, being responsible for the living and their fortunes.[47]

The traditional and cultural instinctive perpetuation of humanity occurs without any forethought regarding population control. The ethics of the African populace are framed by nature and are rooted deeply within the strict confines of indigenous cultural rights. Most Africans believe that having children is a duty advocated by natural phenomena, regardless of planning, affordability, or meeting needs. This perspective contradicts the observation of Ryberg and Tännsjö, who argue for making "the world a better place, not only by making existing creatures happier but also by creating additional happy creatures."[48] Without delving further into the African

---

[44] African Development Bank (AfDB). 2015. "Where Are the Women: Inclusive Boardrooms in Africa?" Accessed December 13, 2023. https://www.afdb.org.

[45] Karima, Cherifa, 2020. "Role of African Women in Development and Economic Life: Reality and Challenges." International Journal of Inspiration & Resilience Economy 4 (1): 1–9, 1.

[46] Martin, Kolk, 2019. "Demographic Theory and Population Ethics—Relationships between Population Size and Population Growth." Working Paper 11. Stockholm: Institute for Futures Studies.

[47] Michael, F.C Bourdillon, 1998. The Shona Peoples: Ethnography of the Contemporary Shona, with Special Reference to Their Religion. Gweru: Mambo Press.

[48] Jesper, Ryberg, and Tannsjo Torbjörn eds, 2004. The Repugnant Conclusion: Essays on Population Ethics. Dordrecht: Kluwer Academic Publishers, 1.

concept of happiness, population ethics assumes that the state of the world, whether better or worse, can depend on the people who are created. To some extent, the way in which Africans perceive procreation may influence parental methods.

Sub-Saharan population ethics are more communitarian and heavily influenced by culture, language, and traditions. The perception of the right to procreation within African indigenous culture is largely seen as a duty, subtly portrayed as an inherent natural function for both males and females. It is, in fact, reasonable to argue that there are no options other than the necessity for all human beings to engage in procreation. Human rights are established through mutual agreement, particularly during social interactions within the community; typically, individuals voluntarily relinquish their rights. African society views life as centered around a 'home' established by a husband, wife, and children.[49] Family serves as the foundation of a comprehensive social life, blossoming into a network of social relationships.[50] Children are regarded as individuals and embodiments of the physical and spiritual future. In Zimbabwe, when someone dies childless, they cannot be accepted into the spirit world; a ritual involving burying the person with a rat is performed to rectify the situation.[51].

Women play a pivotal role in African communities as the backbone of households during times of crisis and development.[52] They are responsible for devising strategies to sustain their families, often working extra hours in formal employment, and thereby challenging the conventional notion that men are solely responsible for providing for their families. Patriarchal systems uphold the prevalence of male dominance and power in African contexts. Boserup underscores the significance of women and their impact on African population ethics, revealing the intricate interconnections between gender, population dynamics, and development.[53]

The treatment of women in respect of infertility reveals that there is limited freedom in choosing to procreate. Infertility is perceived as detrimental to the right to procreate. The World Health Organization's (1987) definition of infertility exhibits an emphasis on women, asserting that infertility occurs when a woman has never conceived, has conceived but cannot conceive again, or has conceived but experienced a stillbirth.[54] It is, therefore, clear that women carry the social burden of infertility. The tendency for women to accept blame for infertility "reflects distinctive virtue whereas men hide behind culture and never accept the blame publicly."[55] Abortion is discouraged. In the context of this study, abortion is not viewed through

[49] John, Locke, 1821. Two Treatises on Government. London: Whitmore and Fenn. Accessed November 1, 2023. https://www.bartleby.com/169/201.html.

[50] Stephen Awoniyi, "African Cultural Values: The Past, Present and Future," Journal of Sustainable Development in Africa 17 (2015): 1–13.

[51] Sande and Takawira-Mtwaya, "Assisted Reproductive Technologies."

[52] Ester, Boserup, 1970. Women's Role in Economic Development. London: Allen & Unwin.

[53] Boserup, Ester. 1970. Women's Role in Economic Development. London: Allen & Unwin.

[54] World Health Organization (WHO). 1987. WHO Laboratory Manual for the Examination of Human Semen and Semen-Cervical Mucus Interaction. 2nd ed. Singapore: Press Concern.

[55] Sande and Takawira-Mtwaya, "Assisted Reproductive Technologies", 34.

the lens of women's health but rather as a means to prevent interference with life. Within African culture, abortion is considered taboo because individuals are driven by the need to manage populations. African traditions uphold the sanctity of human life, positing that society must celebrate life as the primary reason for existence.[56] Cultural paradigms significantly influence the persistent emphasis on banning abortion in most African states. For instance, in Zimbabwe, abortion is illegal and protected under the Termination of Pregnancy Act of 1977 (Chap. 15:10).

5.4.3 *Manifestation of Procreation as a Communal Duty*

The duty of procreation is fascinating in African communities because it is woven into the cultural fabric of many of them. This duty manifests through rituals, social expectations, and economic imperatives. Notable examples include Zimbabwe, where the Shona people regard childlessness as a spiritual issue. Infertility results in one being buried with a rat, symbolizing inferiority, to appease the ancestors for acceptance into the spiritual ancestral realm. Similarly, the Yoruba tribe of Nigeria associates a large family with prosperity, which arises through central approval. Women are pressured to bear children shortly after marriage.[57] Though based on traditional success stories, this expectation naturally conflicts with contemporary goals of female education and economic involvement. How do Yoruba women navigate the demands to prioritize childbearing against opportunities for personal development as urbanization increases? This conflict reveals how collective ethics can stifle individual paths to growth and calls for a reconsideration of 'prosperity' outside of population counts.

The Dagomba in northern Ghana regard polygamy as a strategy to maximize offspring. Bawah & Akweongo explain that a man's social status is linked to the number of children he fathers.[58] Examining this instrumentalization of childbearing for social capital is important: how much does it support gender power inequalities? In contexts where nuclear-family investment, such as school expenses, increasingly determines economic mobility, how can males reconcile conventional status symbols with the current budgetary reality? This highlights a conflict between neoliberal individualism and honor systems based on lineage.

The Maasai of Kenya and those in Tanzania consider marriage to be contingent upon bride wealth payments. A marriage is deemed valid only after childbearing.[59]

---

[56] Farhana Jogee, Is There Room for Religious Ethics in South Africa's Abortion Law? (Durban: University of KwaZulu-Natal, 2018).

[57] Tephen, Awoniyi, 2015. "African Cultural Values: The Past, Present and Future." Journal of Sustainable Development in Africa 17 (1): 1–13.

[58] Ayaga Agula Bawah and Patricia Akweongo, "The Socio-Cultural Context of Family Planning in Rural Northern Ghana," Journal of Health, Population and Nutrition 27 (2009): 537–548.

[59] Ernestina Coast, "Maasai Marriage: A Comparative Study of Kenya and Tanzania," Journal of Eastern African Studies 10 (2016): 283–301.

To a greater extent, these examples illustrate how duty is enacted through spiritual beliefs, marital customs, and communal validation. Here, reproduction serves as the execution of a group contract. How, therefore, can societies combine ancestral validations with state-recognised rights at a time when legal marriage frameworks (e.g., constitutional gender equality provisions) progressively separate the validity of unions from fertility? This dispute draws attention to the jurisdictional struggle over the body between statutory and customary power.

5.4.4 Regional and Cultural Diversity in Procreation Ethics

While emphasizing the influence of indigenous cultures, one must acknowledge the diversity of perspectives within Africa, since not all communities share the same views on procreation. Although cultural norms do dominate across the continent, a new trajectory is emerging where individual reproductive rights are gaining momentum in several African communities. In Rwanda, the work by the Rwandan National Family Planning Policy (2012) has led to a reduction in fertility rates from 6.1 to 4.2 children per woman from 2005 to 2020.[60] The distinct urban and rural divide in Africa necessitates caution against homogenizing procreation ethics. This drop marks a silent revolution: state-approved family planning validates personal decision over group obligation. What cultural trade-offs, nevertheless, take place? How are issues of ancestral continuity handled when policies favor fewer families? The situation in Rwanda shows how top-down rights systems could undermine generational covenants necessitating hybrid ethical norms.

For instance, the Ethiopian Demographic and Health Survey reported that urban women in Ethiopia have an average of 2.6 children, while those living in rural areas have 5.4 children. This difference emphasizes how urbanity drives autonomy. However, the increased fertility of rural women reflects material realities rather than solely 'traditional'—that is, agricultural labor—demands. Analyse this binary: does it represent new forces (e.g., wage economies penalizing motherhood)? Is lower urban fertility really 'empowering'? This blurs the binary oppositions between simple tradition and modernity.

Socio-economic forces further accentuate this urban–rural paradox. Wage systems in metropolitan areas often punish parenting through limited maternity leave and employment discrimination, thus pushing women toward smaller families. Conversely, rural fertility rates remain connected to agricultural livelihoods where children assist with labor-intensive farming and elder care, a functional necessity diminished by land privatization and climate change.[61] These tangible realities

[60] National Institute of Statistics of Rwanda (NISR). 2021. Rwanda Demographic and Health Survey 2019–2020. Kigali: NISR. https://www.statistics.gov.rw/publication/demographic-and-health-survey-20192020.

[61] Renee Pennington and Henry Harpending, The Structure of an African Pastoralist Community: Demography, History, and Ecology of the Ngamiland Herero (Oxford: Clarendon Press, 1993).

reveal how economic survival influences reproductive 'choice,' thereby complicating the dichotomy of 'tradition vs. modernity.'

Family extension and lineages are additional factors to consider. The Akan lineage of Ghana prioritizes the continuation of the lineage; however, this is changing as people increasingly adopt smaller families as a result of economic pressures.[62] This can be contrasted with the Himba people of Namibia, who maintain a high fertility rate, where children are viewed as essential pastoral labor.[63] Land privatization and climate change also pose existential threats to this functionalist perspective on reproduction. How does economic pressure reframe the 'duty' to reproduce when pastoralism becomes unsustainable? The possible decline of this labor–reproduction nexus exposes how the forced reconfiguration of moral demography might result from environmental changes.

Religious influences cannot be undone. Religious diversity shapes attitudes toward procreation.[64] They further explain how in northern Nigeria, where Islam is predominant, polygamy is encouraged, leading to large families, whereas Christian communities, such as those in Nairobi, Kenya, experience slower declines in fertility. These doctrinal differences yield tangible demographic outcomes. Islamic support of polygamy increases pressure on women in northern Nigeria to have multiple children to secure marriage status, often limiting their education and economic opportunities. Conversely, Christian groups in Nairobi show slower fertility declines partly due to religious opposition to contraception; however, rising female education promotes discreet adoption of family planning.[65] This juxtaposition illustrates how gendered socioeconomic pathways interact with theological norms to either reinforce or subtly alter 'obligation.'

The study by Jogee asserts that, in South Africa, the legislature for the Choice on Termination of Pregnancy Act (1996) has granted women autonomy over abortion, and, to a greater extent, this has challenged traditional taboos.[66] Feminist activists have played a pivotal role in promoting individual reproductive rights. For instance, Sylvia Tamale, a feminist scholar, challenges the conflation of womanhood with motherhood and advocates for bodily autonomy.[67] The organization's work, conducted by Ipas Africa, particularly destigmatizes family planning in Nigeria and

---

[62] Baffour K. Takyi and F. Nii-Amoo Dodoo, "Gender, Lineage, and Fertility-Related Outcomes in Ghana," Journal of Marriage and Family 67 (2005): 251–257.

[63] Renee Pennington and Henry Harpending, The Structure of an African Pastoralist Community: Demography, History, and Ecology of the Ngamiland Herero (Oxford: Clarendon Press, 1993).

[64] Thomas E. Dow and Heather E. Reed, "Religious Influences on Fertility in Sub-Saharan Africa: A Comparative Study of Christian and Muslim Communities," Population Studies 69 (2015): 299–315.

[65] Thomas E. Dow and Heather E. Reed, "Religious Influences on Fertility in Sub-Saharan Africa: A Comparative Study of Christian and Muslim Communities," Population Studies 69 (2015): 299–315.

[66] Farhana Jogee, Is There Room for Religious Ethics in South Africa's Abortion Law? (Durban: University of KwaZulu-Natal, 2018).

[67] Sylvia Tamale, "Decolonizing Feminism in Africa: Challenging Cultural and Religious Patriarchies," Feminist Africa 2 (2020): 8–28.

emphasizes informed choice over cultural mandates.[68] Tamale's stance directly contests ancestral paradigms that tether female worth to fertility. However, how do such movements address communal fears of cultural erosion? Feminist critiques must engage with indigenous ontologies (e.g., ancestral continuity as collective survival) to avoid replicating colonial dismissals of 'backwardness.' This demands dialectical ethics, not mere rights imposition.

Feminist interventions are gradually altering conventional wisdom. Access to abortion in South Africa has reduced covert operations and maternal mortality, thereby indirectly decreasing taboos surrounding non-procreative sexuality.[69] Campaigns such as Ipas Africa's destigmatizing of family planning in Nigeria have also assisted women in navigating ancestral expectations, for instance, by spacing births without wholly rejecting motherhood.[70] However, these advances face opposition where concerns about cultural preservation frame rights as 'Western erosion,' thereby emphasizing the necessity for communication with indigenous ontologies (e.g., balancing physical autonomy with ancestral continuity as communal survival).

5.5 Moral Demography as a Solution for African Population Ethics

As it is applied in this chapter, moral demography examines how communal ethics (e.g., ancestral duties, gender standards) and contingent historical circumstances (e.g., colonial policies, religious institutions) co-produce demographic results. Unlike scientific or rights-based approaches, it emphasizes local moral reasoning, revealing why population behaviors (e.g., high fertility) endure despite economic constraints or global standards. Moral demography serves as a solution to African population ethics as it investigates whether, and under what circumstances, one can justify bringing a new life into the world. Individuals coming into existence are harmful because they affect the act of procreation.[71] Parents have a moral obligation to ensure they bring the best children into life and that they may use genetic selection technologies.[72]

Population composition highlights aspects such as where and how people live, providing insights into culture, health, politics, and even economics. According to

[68] Joanna N. Erdman, "Reproductive Rights in Africa: Strategies and Approaches," International Journal of Gynecology & Obstetrics 119 (2011): 198–202.

[69] Farhana Jogee, Is There Room for Religious Ethics in South Africa's Abortion Law? (Durban: University of KwaZulu-Natal, 2018).

[70] Joanna, N, Erdman, 2011. "Reproductive Rights in Africa: Strategies and Approaches." International Journal of Gynecology & Obstetrics 119 (2): 198–202

[71] David Benatar, Better Never to Have Been: The Harm of Coming into Existence (Oxford: Oxford University Press, 2006).

[72] Julian Savulescu, "Procreative Beneficence: Why We Should Select the Best Children," Bioethics 15 (2001): 413–426.

the World Population Prospects (2015), fertility, mortality, and migration status assist in estimating population growth.[73] This article employs moral demography as a theoretical lens to understand the complexities of reconstructing the population of Africa. Originally applied to English peasants who protested unfair grain prices, Edward Thompson's idea of 'moral economy'—based on communal concepts of justice rather than mere market logic—serves as the foundation for the development of moral demography.[74] Transposed to demography, this approach elucidates how culturally specific ethics, such as ancestral obligations in Ghanaian Akan society, rather than individual cost–benefit assessments, shape reproductive decisions (e.g., high fertility). Sayer highlights the difference: whereas moral demography investigates how moral rules (e.g., lineage honor) combine with historical causes (e.g., colonization) to influence population outcomes, moral economy studies economic behaviors constrained by social conventions.[75]

Reconstructing the demography of Africa is notoriously difficult due to the lack of sources, and moral demography employs a selective approach to understand demographic patterns. According to Johnson-Hanks, "In some cases, selectivity effects are not threats to the data quality, but are instead the processes most central to understanding a phenomenon;' it allows us to take 'structure' seriously, guided by the voices of those experiencing and enacting change."[76] Moral demography suggests that within communities, various factors, such as colonization and Christianity, play a significant role; one must comprehend the processes of both mutuality and exchange.[77] Other potential selective elements include land, social justice, and moral ethnicities, which are critical to African history.[78] Guided by moral demography, this chapter focuses on the right to procreation. Broadly, the right to procreation falls within the framework of reproduction. The concept of reproduction in an African context oscillates between traditional values and adapting to changing circumstances.[79] In addition to prioritizing means of production, moral demography incorporates other aspects of the family that are critical to this study because the right to procreate is the bedrock of most African families.

---

[73] World Population Prospects. 2015. Accessed December 23, 2023. http://esa.un.org/unpd/wpp.

[74] Edward P. Thompson, 1971. "The Moral Economy of the English Crowd in the Eighteenth Century." Past & Present 50 (1): 76–136.

[75] Andrew Sayer, "Moral Economy and Political Economy," Studies in Political Economy 61 (2000): 79–104.

[76] Jennifer Johnson-Hanks, "Uncertainty and the Second Space: Modern Birth Timing and the Dilemma of Education," European Journal of Population 20 (2004): 351–373, 320.

[77] Jennifer Johnson-Hanks, "Uncertainty and the Second Space: Modern Birth Timing and the Dilemma of Education," European Journal of Population 20 (2004): 351–373.

[78] John, Lonsdale, 1992. "The Moral Economy of Mau Mau." In Unhappy Valley: Conflict in Kenya and Africa, Book 2: Violence and Ethnicity, edited by Bruce Berman and John Lonsdale, 265–468. London: James Currey.

[79] Thomas, Spear, 1997. Mountain Farmers: Moral Economies of Land and Agricultural Development in Arusha and Meru. Berkeley: University of California Press, 11–12.

One example of utilizing elements of moral demography is employing religious registers to establish African population demographics. Walters used missionary intimate records to count, determine, and monitor souls by measuring mortality, fidelity, and faith.[80] Accordingly, Walter argues that.

> This intimate recording means that the African parish registers, together with related sources such as missionary diaries and letters and oral histories, describe qualitatively and quantitatively what happens to individual agency (reproductive decision-making) when the moral hegemony shifts (via evangelization and colonization), and how the two interact in a reciprocal process of change.[81]

In other words, the missionary registers provided new opportunities for moral demography, utilizing the fertility transitions in Africa.[82] Using Christian registers has the advantage of offering accurate demography, as most employment arose directly from the church. After all, the church in Africa was instrumental in education, with many Christian leaders dominating the lower ranks of colonial administration and later using these positions to lead the nationalistic independence movement.[83] Furthermore, the church, as an institution, had access to health and education services and collaborated with the Catholic moral order to instil the normative Western model of family, domestic life, sexuality, and even control (Hunt[84]; Smythe[85]). Christianity defines moral discourse within sexual relationships.[86]

5.6 Conclusion

This chapter examines African procreative ethics through a moral and cultural lens, emphasizing that in many African communities, the act of procreation is not simply a personal choice, but a communal obligation embedded in religious beliefs, ancestral practices, and socio-economic expectations. Across diverse regions, from the Shona in Zimbabwe to the Himba in Namibia, procreation serves as a vehicle for maintaining lineage, fulfilling spiritual duties, and asserting social identity. However,

[80] Sarah Walter, "Counting Souls: Towards an Historical Demography of Africa," Demographic Research 34 (2016).

[81] Sarah Walter, "Counting Souls: Towards an Historical Demography of Africa," Demographic Research 34 (2016): 63.

[82] Sarah Walter, "Counting Souls: Towards an Historical Demography of Africa," Demographic Research 34 (2016): 63.

[83] John McCracken, 2008. Politics and Christianity in Malawi, 1875–1940: The Impact of the Livingstonia Mission in the Northern Province. 3rd ed. Zomba: Kachere Series.

[84] Nancy Rose, Hunt 1999. A Colonial Lexicon: Of Birth Ritual, Medicalization, and Mobility in Congo. Durham: Duke University Press.

[85] Kathleen R. Smythe, 2006. Fipa Families: Reproduction and Catholic Evangelization in Nkansi, Ufipa, 1880–1960. Portsmouth, NH: Heinemann.

[86] Sarah Walter, "Counting Souls: Towards an Historical Demography of Africa," Demographic Research 34 (2016): 93.

Africa is not culturally monolithic. As illustrated by contemporary shifts in reproductive behaviors in countries like Rwanda, Ethiopia, and South Africa, new trajectories are emerging. These reflect increasing engagement with reproductive rights, economic realities, and feminist critique, highlighting a growing tension between inherited moral codes and the pursuit of individual autonomy. The chapter, therefore, calls for an ethical reading that respects both continuity and change in African fertility cultures.

By employing moral demography as an interpretive framework, this chapter offers a tool for comprehending African population dynamics. Unlike purely statistical or biomedical models, moral demography captures the interaction of moral codes, historical legacies, and evolving social values to shape demographic outcomes. It enables us to inquire not only how populations change, but also why and under whose moral authority. The chapter posits that the future of African population ethics lies in maintaining a productive tension between the values of community and individuality, as well as tradition and transformation. Ethical approaches attentive to African contexts must refrain from imposing universalist assumptions and instead engage with the lived realities, narratives, and moral economies that continue to shape how Africans navigate the rights and duties associated with procreation.

References

African Development Bank. 2015. *Where are the women: Inclusive boardrooms in Africa?* Accessed December 13, 2023. https://www.afdb.org.

African Union. 1986. *African charter on human and peoples' rights*. OAU Doc. CAB/LEG/67/3 rev. 5. https://au.int/en/treaties/african-charter-human-and-peoples-rights.

Anaya, James. 2004. *Indigenous peoples in international law*. Oxford: Oxford University Press.

Anaya, James, and Robert Williams. 2001. *The protection of indigenous peoples' rights over lands and natural resources under the inter-American human rights system*. Harvard Human Rights Journal. Accessed December 22, 2023. https://scholar.law.colorado.edu/faculty-articles/846

Arrhenius, Gustaf, Krister Bykvist, Tim Campbell, and Elizabeth Finneron-Burns. 2022. Introduction. In *The Oxford handbook of population ethics*, 1–12. Oxford: Oxford University Press.

Awoniyi, Stephen. 2015. African cultural values: The past, present and future. *Journal of Sustainable Development in Africa* 17 (1): 1–13.

Barume, Albert Kwokwo. 2009. Responding to the concerns of the African states. In *Making the declaration work: The United Nations declaration on the rights of indigenous people*, ed. Claire Charters and Rodolfo Stavenhagen. Copenhagen: IWGIA.

Bawah, Ayaga Agula, and Patricia Akweongo. 2009. The socio-cultural context of family planning in rural Northern Ghana. *Journal of Health, Population and Nutrition* 27 (4): 537–548.

Benatar, David. 2006. *Better never to have been: The harm of coming into existence*. Oxford: Oxford University Press.

Bourdillon, Michael F. C. 1998. *The Shona peoples: Ethnography of the contemporary Shona, with special reference to their religion*. Gweru: Mambo Press.

Boserup, Ester. 1970. *Women's role in economic development*. London: Allen & Unwin.

Cleland, John, and Kazuyo Machiyama. 2017. The challenges posed by demographic change in sub-Saharan Africa: A concise overview. *Population and Development Review* 43 (S1): 264–286.

Cherifa, Karima. 2020. Role of African women in development and economic life: Reality and challenges. *International Journal of Inspiration & Resilience Economy* 4 (1): 1–9.

Coast, Ernestina. 2006. Maasai marriage: A comparative study of Kenya and Tanzania. *Journal of Eastern African Studies* 10 (2): 283–301. https://doi.org/10.1080/17531055.2016.1181412.

Del R. Vincent. 2023. 10 reasons why people want kids (and ten reasons they Don't). Accessed November 3, 2023. https://wehavekids.com/having-baby/Most-Common-Reasons-Why-People-Want-Children.

Doyle, Shane. 2013. Demography and disease. In *The Oxford handbook of modern African history*, ed. John Parker and Richard Reid, 38–55. Oxford: Oxford University Press.

Dow, Thomas E., and Heather E. Reed. 2015. Religious influences on fertility in sub-Saharan Africa: A comparative study of Christian and Muslim communities. *Population Studies* 69 (3): 299–315.

Erdman, Joanna N. 2011. Reproductive rights in Africa: Strategies and approaches. *International Journal of Gynecology & Obstetrics* 119 (2): 198–202.

Ethiopian Central Statistical Agency (CSA). 2016. *Ethiopian demographic and health survey 2016*. Addis Ababa: CSA. https://dhsprogram.com/pubs/pdf/FR328/FR328.pdf.

Eze, Emmanuel Chukwudi, ed. 1997. *African philosophy: An anthology*. Oxford: Blackwell Publishers.

Foucault, Michel. 2018. In *Histoire de la sexualité 4: Les aveux de la chair*, ed. Frédéric Gros. New York: Gallimard.

Gilbert, Jérémie. 2009. Custodians of the land: Indigenous peoples, human rights and cultural integrity. In *Cultural diversity, heritage and human rights*, ed. Michelle Langfield, William Logan, and Máiréad Nic Craith, 31–44. London: Routledge.

Hopwood, Nick, Rebecca Flemming, and Lauren Kassell, eds. 2018. *Reproduction: Antiquity to the present day*. Cambridge: Cambridge University Press.

Finkle, Jason L., and C. Alison McIntosh. 1994. The new politics of population: Conflict and consensus in family planning. *Population and Development Review* 20 (Supplement): 3–38.

Jogee, Farhana. 2018. *Is there room for religious ethics in South Africa's abortion law?* Durban: University of KwaZulu-Natal.

Johnson-Hanks, Jennifer. 2004. Uncertainty and the second space: Modern birth timing and the dilemma of education. *European Journal of Population* 20 (4): 351–373.

Kjekshus, Helge. 1996. *Ecology control and economic development in East African history: The case of Tanganyika 1850–1950*. 2nd ed. London: James Currey.

Kolk, Martin. 2019. Demographic theory and population ethics—relationships between population size and population growth. In *Working Paper 11*. Stockholm: Institute for Futures Studies.

Locke, John. 1821. *Two treatises on government*. London: Whitmore and Fenn. Accessed November 1, 2023. https://www.bartleby.com/169/201.html.

Lonsdale, John. 1992. The moral economy of Mau Mau. In *Unhappy valley: Conflict in Kenya and Africa, book 2: Violence and ethnicity*, ed. Bruce Berman and John Lonsdale, 265–468. London: James Currey.

Lovejoy, Paul. 2011. *Transformations in slavery: A history of slavery in Africa*. 3rd ed. Cambridge: Cambridge University Press.

Matumbu, Faith. 2023. Women's rights: A precursor for African development. In *Contemporary development ethics from an African perspective*, ed. Beatrice Okyere-Manu, Stephen Nkansah Morgan, and Obvious Katsaura, 157–172. Cham: Springer.

Metz, Thaddeus. 2014. Ubuntu and the debate on African philosophy. *African Studies Review* 57 (2): 21–39.

Mbacké, Cheikh. 2017. The persistence of high fertility in sub-Saharan Africa: A comment. *Population and Development Review* 43 (S1): 330–337.

McCracken, John. 2008. *Politics and Christianity in Malawi, 1875–1940: The Impact of the Livingstonia Mission in the Northern Province*. 3rd ed. Zomba: Kachere Series.

National Institute of Statistics of Rwanda (NISR). 2021. *Rwanda demographic and health survey 2019–2020*. Kigali: NISR. https://www.statistics.gov.rw/publication/demographic-and-health-survey-20192020.

Pennington, Renee, and Henry Harpending. 1993. *The structure of an African pastoralist community: Demography, history, and ecology of the Ngamiland Herero*. Oxford: Clarendon Press.
Ryberg, Jesper, and Torbjörn Tännsjö, eds. 2004. *The repugnant conclusion: Essays on population ethics*. Dordrecht: Kluwer Academic Publishers.
Sande, Nomatter, and Tawanda Takawira-Mtwaya. 2022. Assisted reproductive technologies, infertility and women in Zimbabwe. In *Religion, women's health rights, and sustainable Development in Zimbabwe*, ed. Sophia Chirongoma, Molly Manyonganise, and Ezra Chitando, 45–62. Cham: Springer.
Sande, Nomatter, and Tawanda Takawira-Mtwaya. 2023. Cultural stigma and infertility in contemporary Zimbabwe. *Journal of African Cultural Studies* 35 (1): 28–42.
Sasser, Jade. 2018. *On infertile ground: Population control and women's rights in the era of climate change*. New York: New York University Press.
Savulescu, Julian. 2001. Procreative beneficence: Why we should select the best children. *Bioethics* 15 (5–6): 413–426.
Sayer, Andrew. 2000. Moral economy and political economy. *Studies in Political Economy* 61 (1): 79–104.
Scott, James C. 1977. *The moral economy of the peasant: Rebellion and subsistence in Southeast Asia*. New Haven: Yale University Press.
Schumacher, Bastian. 2019. Stalls in fertility transitions in sub-Saharan Africa: Revisiting the evidence. *Studies in Family Planning* 50 (3): 257–258.
Sibanda, Amson, Zerihun Woubalem, Donald P. Hogan, and David P. Lindstrom. 2003. The proximate determinants of the decline to below-replacement fertility in Addis Ababa, Ethiopia. *Studies in Family Planning* 34 (1): 1–7.
Smythe, Kathleen R. 2006. *Fipa families: Reproduction and Catholic evangelization in Nkansi, Ufipa, 1880–1960*. Portsmouth, NH: Heinemann.
Spear, Thomas. 1997. *Mountain farmers: Moral economies of land and agricultural development in Arusha and Meru*. Berkeley: University of California Press.
Strathern, Marilyn. 1992a. *After nature: English kinship in the late twentieth century*. Cambridge: Cambridge University Press.
Strathern, Marilyn. 1992b. *Reproducing the future: Anthropology, kinship, and the new reproductive technologies*. Manchester: Manchester University Press.
Strathern, Marilyn. 1993. Postscript: A relational view. In *Technologies of procreation: Kinship in the age of assisted conception*, ed. Jeanette Edwards et al., 2nd ed., 153–175. London: Routledge.
Takyi, Baffour K., and F. Nii-Amoo Dodoo. 2005. Gender, lineage, and fertility-related outcomes in Ghana. *Journal of Marriage and Family* 67 (1): 251–257.
Tamale, Sylvia. 2020. Decolonizing feminism in Africa: Challenging cultural and religious patriarchies. *Feminist Africa* 2 (1): 8–28.
Thompson, Edward P. 1971. The moral economy of the English crowd in the eighteenth century. *Past & Present* 50 (1): 76–136.
United Nations. 2019. *World population prospects 2019: Highlights*. New York: United Nations Department of Economic and Social Affairs.
Walter, Sarah. 2016. Counting souls: Towards an historical demography of Africa. *Demographic Research* 34 (3): 63–108. Accessed December 21, 2023. http://www.demographic-research.org/Volumes/Vol34/3/.
Weber, Hildegard, Jens Dahl, Fiona Wilson, and Espen Waehle. 1993. *Never drink from the same cup: Proceedings of the conference on indigenous peoples in Africa*. Copenhagen: IWGIA.
Wiredu, Kwasi. 2020. The African concept of personhood. In *The Stanford encyclopedia of philosophy*, edited by Edward N. Zalta. https://plato.stanford.edu/entries/african-personhood/.
World Health Organization. 1987. *WHO Laboratory manual for the examination of human semen and semen-cervical mucus interaction*. 2nd ed. Singapore: Press Concern.
World Population Prospects 2015. Accessed December 23, 2023. http://esa.un.org/unpd/wpp.
Yamada, Akira. n.d. *Population and development in Africa: The role of indigenous knowledge*. Tokyo: United Nations University Press.

Chapter 6
Rethinking Procreation in the Era of Climate Crisis: An Ethico-Cultural Critique

Hupenyu Makusha

Abstract This chapter examines the ethical dilemmas associated with procreation in the context of escalating climate change, with a particular focus on Zimbabwe. It interrogates the moral implications of bringing new life into a world increasingly influenced by environmental degradation, economic hardship, and social instability. The study investigates the justifiability of having children when future generations are expected to endure the consequences of climate-related disasters. Through the analytical lens of African communitarianism, which emphasizes collective responsibility and intergenerational obligations, this chapter posits that the moral question of procreation must be re-evaluated within the extensive community resilience framework and environmental ethics in Africa, and specifically Zimbabwe. While some people may advocate for restraint in reproduction owing to the suffering climate change is anticipated to inflict, African ethical traditions suggest that hope and communal action can serve as a foundation for mitigating harm and ensuring a sustainable future for children. Ultimately, this chapter contends that moral responsibility transcends individual choices and extends to collective environmental stewardship, urging communities to embrace sustainable practices that protect both present and future generations.

Keywords Ethics · Procreation · Climate change · African communitarianism · Intergenerational

6.1 Introduction

Climate change presents significant ethical challenges, where environmental degradation intensifies socio-economic vulnerabilities in Africa, particularly in Zimbabwe. The deterioration of agricultural productivity, food insecurity, and

---

H. Makusha (✉)
University of Kwazulu-Natal, Ethics, Pietermaritzburg, South Africa

B. Okyere-Manu, P. Mwipikeni (eds.), *An African Approach to Population Ethics*, Advancing Global Bioethics 23,
https://doi.org/10.1007/978-3-032-24464-2_6

health risks disproportionately affects marginalized communities, raising ethical concerns regarding intergenerational responsibility. The intersection of climate change, cultural traditions, and procreation necessitates a critical examination of the moral implications of bringing new life into an uncertain future.

In numerous African cultures, such as Shona, procreation sustains familial lineage and cultural heritage. However, questions arise concerning the ethical justification for reproduction and the associated responsibilities in the context of climatic adversity. This chapter investigates procreative ethics through the lens of African communitarianism, which emphasizes collective responsibility and obligations to future generations. By prioritizing community welfare, this framework offers a vital perspective on the moral dilemmas of procreation in a climate crisis. Ultimately, this analysis contributes to the discourse on climate ethics and procreative justice, calling for a reassessment of the implications of reproduction in an increasingly precarious world.

6.2 The Effect of Climate Change in Zimbabwe

At present, Zimbabwe is undergoing substantial climatic transformations that have extensive repercussions for both its environment and populace. Among these transformations are: extended periods of drought, which severely reduce water availability and agricultural productivity; irregular precipitation patterns that disrupt the cycles of planting and harvesting; increasing temperatures that intensify pre-existing environmental challenges; and catastrophic floods that can devastate crops and livelihoods.[1]These climatic phenomena are not merely isolated occurrences; they have collectively imposed substantial challenges on agriculture and the overall well-being of the nation's population. This highlights the pressing necessity for adaptive strategies in agricultural practices to alleviate the detrimental impacts of climate change and secure food availability for all citizens.

Adverse environmental challenges go beyond immediate agricultural losses, threatening the fundamental aspects of food security and triggering economic instability. Social unrest emerges when communities grapple with heightened poverty and marginalization. Such circumstances elicit significant ethical inquiries on the sustainability of future conditions for succeeding generations who may inherit a landscape characterized by scarcity and uncertainty.

Rural communities such as those in Zimbabwe, which depend on subsistence farming as their primary means of survival, find themselves in a precarious position. These communities experience heightened vulnerability due to the depletion of natural resources, including soil fertility and water sources, and the increasing

[1] Tafadzwa, Chigodora. 2019. 'Climate change and Rural Livelihoods In Zimbabwe: A socioeconomic Analysis'. (Journal of African Studies, 12(2)). pp. 43

unpredictability of weather patterns.[2] As these alterations persist, these communities' challenges will affect them directly and reverberate throughout Zimbabwe's broader socioeconomic framework, necessitating urgent attention and action.[3] It is imperative to acknowledge that the consequences of these challenges extend beyond the local context, influencing national dynamics and requiring a collective response to address the urgent issues at hand. The following is a reflection upon some of the calamities Zimbabwe has encountered, which may have prompted a reconsideration of reproductive behaviors.

6.3 Climate Disasters in Zimbabwe

Zimbabwe has faced numerous catastrophic climate-related disasters in recent decades, which have impacted the country's population in various ways, including starvation, floods, and the destruction of property.

A severe drought in the region, from 2002 to 2003, had a profound and lasting impact on agricultural production, resulting in repercussions far beyond the immediate effects on farming.[4] This environmental crisis led to widespread food shortages, affecting nearly 7.2 million people, and triggering a humanitarian crisis that demanded urgent attention. The situation, as reported by the Food and Agriculture Organization (FAO) in 2004, underscores the extensive implications of droughts on food security and the livelihoods of numerous communities that depend on agriculture for survival.

The inadequate rainfall, which has become increasingly prevalent due to changing climate patterns, has posed significant challenges for farmers across the region. Crop yields have plummeted, diminishing the availability of essential staple foods and driving up prices. Consequently, families found it increasingly challenging to access nutritious food, resulting in a heightened vulnerability to hunger and malnutrition among the impacted populations.

[5]The FAO's findings highlight the immediate consequences of drought on agricultural output and the long-term social and economic ramifications threatening the stability and well-being of entire communities.

---

[2]Tafadzwa, Chigodora. (1999) "*Climate Change and Rural Livelihood In Zimbabwe: A Socioeconomic Analysis*". (Journal of African Studies, 12(2)). 88.

[3]Robert, Smith. (2020) "*Cultural Expectations in Modern Society, Culture and Society Review*". 12, no. 1: 1–15.

[4]Food and Agriculture Organisation of the United Nations (FAO), *The State of Food Security and Nutrition in the World 2022* (Rome: FAO, 2022).

World Bank, *World Development Report 2021: Data for Better Lives* (Washington, DC: World Bank, 2021).

[5]Food and Agriculture Organisation of the United Nations (FAO), *The State of Food Insecurity in the World 2004* (Rome: FAO, 2004), 28.

A severe drought in 2015–2016 devastated Zimbabwe's agriculture, leading to widespread crop failures and animal starvation across the region. As a result of this natural disaster, more than four million people in Zimbabwe were in urgent need of food assistance, highlighting the critical humanitarian crisis that the country faced during this period.[6]According to a report by the United Nations Development Programme, the impact of the drought not only threatened the livelihoods of many farmers but also exacerbated food insecurity for millions of residents of drought-affected Zimbabwe, which required an urgent need for effective intervention and support strategies to alleviate the suffering caused by such climatic conditions.

Then there was Cyclone Idai, which hit in 2019 and is remembered as one of the most devastating tropical cyclones to affect the country. It claimed over 340 lives, profoundly impacting many families and their futures, including their ability to stay together and nurture children amidst the devastation.[7]The extensive destruction of infrastructure and the displacement of numerous families in the worst-hit areas of Chimanimani and Chipinge only added to these challenges. This situation underscores the significant impact of natural disasters on families, highlighting the urgent need for effective disaster response and support systems to safeguard vulnerable communities. The fallout from the storm illustrates the severe consequences, emphasizing how it damaged homes and buildings and disrupted essential services, resulting in a long-lasting humanitarian crisis. This crisis profoundly affects family stability, making it increasingly difficult for those in the affected regions to contemplate having and raising children in such an unstable environment. This draws attention to a broader social issue, revealing that natural disasters have immediate effects and reshape communities' demographic and psychological landscape.

In 2021, heavy rains caused significant flooding in low-lying areas in Zimbabwe, such as Muzarabani and Tsholotsho, overwhelming drainage systems and exacerbating the situation. As noted by some residents, the floods came unexpectedly, taking away everything they had. The disaster displaced thousands of people, uprooting families who lost homes and belongings, such as cattle, sheep, etc. Many sought refuge in makeshift shelters or relied on relatives.[8]Community leader John Dube remarked, "It is heartbreaking to see families scattered and in need of necessities." As communities deal with the aftermath, they struggle to find shelter, food, and emotional support. Destruction has changed living conditions, prompting leaders and organizations to assist where possible.[9]As the Zimbabwe Red Cross outlined, "We are doing our best to provide relief to those affected, but it is a massive challenge." The long-term implications for families and communities are signifi-

---

[6] United Nations Development Programme (UNDP), *Human Development Report 2016: Human Development for Everyone* (New York: UNDP, 2017), 65.

[7] Trust, Mavhura, (2010). "*Climate Change Adaptation in Rural Zimbabwe: Community Initiatives and Institutional Support*". (Cham: Palgrave Macmillan, 2020), 57. Ednah, Chibamu. (2020) Youth and Livelihoods in Zimbabwe: Challenges and Prospects (Harare: Weaver Press). United Nations, World Population Prospects 2019: Highlights (New York: United Nations).

[8] John, Dube. (2023). Personal interview.

[9] Zimbabwe Red Cross Society, *Annual Report 2023* (Harare: Zimbabwe Red Cross Society, 2023)

cant, affecting relationships and resource availability. Supporting families is essential as they adapt to these changes. The question is: To what extent did these disasters influence procreation?

Children died, like many other people.[10]They were also affected in their learning programs, as floods made it impossible to travel to schools.[11]This educational disruption severely hindered many children's academic progress and emotional stability. Starvation mainly affects children, leading to malnutrition. This illustrates how crises disproportionately affect vulnerable groups such as children, compromising their health and development.[12]During this crisis, people continued to procreate, as evidenced in the section on population growth. This shows that despite adversity, population dynamics remained constant, further straining limited resources and social services.

Additionally, tradition and culture significantly influence procreation practices in Zimbabwe, which are shaped by societal expectations, religious beliefs, and customary laws. Below are some key aspects:

6.4 Cultural Expectations and Family Continuity

[13]In many African communities, like those in Zimbabwe, having children is viewed not merely as a personal decision but also as a crucial element of cultural continuity and a way to preserve family lineage.[14] "It is through the rituals, rites and ceremonies that the couples receive blessings from the ancestors and the community as a whole." This highlights the importance of cultural practices in strengthening the bonds between individuals and their heritage.[15]Procreation is deeply ingrained in societal values and often perceived as a sacred duty transcending individual desires. This perspective highlights the collective responsibility individuals feel toward family and community continuity.[16]Childbearing is highly prized in these communities, particularly the birth of male offspring. This cultural emphasis highlights the

[10]United Nations Children's Fund (UNICEF), *Education Disrupted: The Impact of Emergencies on Children's Learning* (New York: UNICEF, 2022), https://www.unicef.org, p 15

[11]World Health Organisation (WHO), *Malnutrition in Children: Key Facts and Global Response* (Geneva: WHO, 2023), https://www.who.int.p 8

[12]World Bank, *Poverty and Shared Prosperity 2022: Correcting Course* (Washington, DC: World Bank, 2022), 22.

[13]Anesu, Chakaipa. (2017). "Cultural Perspectives on Family Structures in Zimbabwe," *Journal of African Studies*.

[14]Beatrice, Okyere-Manu. (2015). "Cohabitation in Akan Culture of Ghana: An Ethical Challenge to Gatekeepers of Indigenous Knowledge System in the Akan Culture," *Alternation Special Edition* 14: 49.

[15]Charles, Mavhunga. (2019). "The Sacred Duty of Procreation: Family and Cultural Preservation," (*Zimbabwean Journal of Social Sciences*) no. 1: 58–75.

[16]Daisy, Dube and Milton, Chimonyo. (2018). "Gender Roles and Childbearing in Zimbabwe,(Journal *of African Studies*), no. 2: 134–150.

importance of male lineage and the continuation of family names within the community.[17]Sons are not only regarded as the primary heirs who will carry on the family name; they also assume vital roles in inheriting property and managing family assets, thereby ensuring the economic stability and legacy of the family unit.[18]The emphasis on having sons also reflects traditional beliefs about gender roles and the expectations placed on individuals within the community, further illustrating the complexities of family dynamics in these societies. This reinforces the notion that societal values have a significant influence on personal and familial relationships.

6.5 Marriage and Procreation

[19]In traditional Zimbabwean society, marriage holds profound significance beyond mere cohabitation; it is fundamentally tied to procreation and the expansion of families.[20]Okyere-Manu alludes to this by saying, "This is because they believe that the survival of the community depends on the institution of marriage and procreation attached to it." In this context, couples, particularly women, face societal expectations to begin having children soon after they enter marriage.[21]This expectation illustrates that a woman's primary role is often viewed as a bearer of children, reflecting traditional attitudes toward gender roles and family structure. This perspective can limit women's opportunities for personal and professional growth, reinforcing cultural stereotypes.

[22]Furthermore, the concept of bride price, known as lobola, is a tangible representation of a family's investment in a woman's reproductive potential. By paying lobola, a man's family acknowledges the value of a woman not only as a partner but also as a future mother.[23]This practice reinforces the belief that marriage and procreation are inherently intertwined aspects of life in this culture, where the continuation of lineage and nurturing future generations is paramount. The dynamics of marriage and the expectations surrounding procreation in Zimbabwean society are significantly influenced by deep-rooted cultural beliefs and practices that shape family and social structures. However, climate change disasters increasingly

[17] Never, Mujere. (2021). "Economics of Family Legacy in Zimbabwe"(*Journal of Economic and Social Research*) no. 3 (2021): 200–215.

[18] Loveness, Sibanda. (2022). *Gender Expectations and Family Dynamics in Zimbabwe* (Harare: University of Zimbabwe Press) 70.

[19] Rumbidzai, Chireshe. (2011). "*Cultural Practices Influencing Women's Roles in Zimbabwean Society*". (Harare: Zimbabwe University Press, 2011), 23.

[20] Okyere-Manu. 2015. Cohabitation in … 48

[21] Alois, Mlambo. (2010). "*Gender Roles and Marriage in Zimbabwe*". (Bulawayo: Books of Zimbabwe, 2014). 17.

[22] Ndlovu. 2010. Lobola; … 32

[23] Khabele, Matlosa. (2015). "Marriage, Family Structure, and Social Expectations in Zimbabwe," (*African Studies Review* 58, no. 2 (2015)): 22.

challenge these traditional norms, which disrupt communities' natural environment and social fabric.

As families face the consequences of extreme weather events, such as droughts or floods, their ability to uphold cultural practices regarding marriage and child nurturing becomes strained. This intersection of cultural norms and climate change illustrates how environmental challenges can alter human relationships and societal expectations, urging a re-evaluation of traditional roles in emerging ecological realities.

6.6 The Influence of the Extended Family and Elders in Procreation

The extended family plays a crucial role in procreation, particularly in cultures where familial ties are paramount.[24]In many African societies, if a couple struggles to conceive, there may be significant pressure from elders to seek traditional remedies or even consider polygamy as a solution. This cultural expectation often leads couples to explore various options outside of medical intervention.

In cases of infertility, families often suggest alternative solutions that align with traditional practices. One common practice is the husband's brother stepping in, a practice known as levirate marriage, which is prevalent in certain cultures in Zimbabwe.[25]This involves the brother marrying the widow to ensure the continuation of the family line.[26]Additionally, a woman may be encouraged to bear children with another male relative, which can reinforce family bonds and uphold societal expectations regarding procreation. Such practices highlight the significant impact of family dynamics and cultural traditions on reproductive choices.

[24] Linda, Smith. (2020). "Urbanisation and Its Effect on Cultural Values," *Urban Sociology* 12, no. 2 (2020): 89–102.

[25] Angela, Jones. (2019) "Cultural Practices in Family Dynamics: The Role of Levirate Marriage," (*Journal of Ethnographic Studies* 12, no. 3 (2019)): 68.

[26] Mary, Taylor. (2021). "The Interplay of Family and Fertility: Alternative Solutions to Infertility,"(*Family Relations* 70, no. 4 (2021)): 599–612.

6.7 Religious and Spiritual Beliefs Surrounding Procreation

[27]Traditional African spirituality and Christianity significantly shape perceptions of fertility within various communities.[28]In many cultures, children are viewed not merely as a personal blessing but as a vital link to ancestors and a divine gift from God. This perspective highlights children's profound spiritual and cultural significance within these societies, reinforcing that their existence interweaves the past, present, and future generations.[29]This deep-seated belief underscores a cultural narrative where the presence of children is intertwined with familial honor and spiritual fulfilment. This reflects a broader societal expectation that emphasizes the importance of family and the role of children in contributing to one's legacy and identity.

[30]Conversely, infertility is often seen as a profound social and spiritual concern. This highlights the emotional weight that infertility carries, impacting individuals and their communities on multiple levels.[31]Many individuals attribute their struggles with conceiving to various spiritual causes. This perspective highlights the complex interplay between spirituality and personal health journeys.[32]These can include the influence of witchcraft, the effects of curses, or even the perception of divine punishment for past misdeeds. These elements can significantly shape cultural beliefs and practices, influencing how individuals interpret their experiences.[33]Thus, the interplay between cultural beliefs and fertility underscores a complex landscape where spirituality and medicine often intersect, highlighting the rich nature of health and well-being in these communities. This interaction shapes individual experiences and influences broader societal norms regarding health practices.

[27] Olufunke, Ogunyemi. (2007). "African Cultures and Fertility" (Journal of Cultural Perspectives, 2007). 46

[28] Moyo, R. (2012). The spiritual dimensions of fertility in traditional settings. (African Journal of Spirituality). 28.

[29] John, Smith. (2015). "Familial Honour and Childbearing: A Cultural Analysis" (*Sociology of Family Studies*). 113

[30] Angela, Jones. (2019). "Cultural Practices in Family Dynamics: The Role of Levirate Marriage," (*Journal of Ethnographic Studies* 12, no. 3) 56.

[31] Patricia, Nichols. (2018). "Beliefs and Practices Surrounding Infertility" (*Journal of Cross-Cultural Research*). 42.

[32] Tunde, Adesina. (2014). "Witchcraft, Curses, and Fertility: A Cultural Assessment" (*Journal of Anthropology and Culture*). 77.

[33] Adebayo, Elijah. (2019). "Intersecting Paths: Spirituality and Medicine in African Communities" (*Journal of Health and Spirituality*). 91

6.8 Polygamy and Procreation

[34]In certain traditional societies in Zimbabwe, the practice of polygamy continues to be prevalent, influenced by various cultural and social factors. This practice consistently shapes familial structures and societal norms within these communities.[35]A significant motivation for this practice is the aspiration to have large families, which is often regarded as a symbol of prosperity and success within these communities. This perspective reflects entrenched cultural values prioritizing family size as a measure of social status and community strength. Within these societies, men who maintain multiple wives generally find that they can sire numerous children.[36]This reinforces the man's social status and bolsters economic security.[37]A more prominent family can provide additional hands to aid with agricultural tasks or other family enterprises, enhancing the family's overall productivity and resource management. This augmented collaboration improves efficiency and fortifies familial bonds and community engagement.

[38]Furthermore, having multiple children serves as a strategy to ensure that some individuals survive into adulthood, which holds particular significance in contexts where child mortality rates may be elevated.[39]Consequently, polygamy can be perceived not merely as a personal choice but also as a notable cultural practice that intersects with issues of social standing, economic stability, and familial cooperation. This practice frequently mirrors deeply entrenched societal norms and values that shape individuals' identities and relationships.

6.9 Women's Roles and Reproductive Responsibility

[40]Women frequently encounter societal pressure and scrutiny about their fertility, as they are often held accountable for reproductive outcomes. This expectation can result in considerable stress and anxiety, adversely affecting their overall

---

[34] Godfrey, Mavengere. (2018). "Polygamy and Its Impact on Society" (*Zimbabwean Journal of Sociology*). 178.

[35] Anesu, Chakaipa. (2017). "Cultural Perspectives on Family Structures in Zimbabwe" (*Journal of African Studies*). 46

[36] Vimbai, Mudzamiri. (2020). "Socioeconomic Factors Influencing Polygamous Marriages" (*African Journal of Family Studies*). 112

[37] Rufaro, Ndlovu. (2019). "Agricultural Practices and Family Labour in Zimbabwe" (*Zimbabwean Agricultural Review*). 72

[38] Tendai, Munsaka. (2021) "Child Mortality and Family Dynamics in Zimbabwe,"(*International Journal of Health Studies)*. 58

[39] Blessing, Mudavanhu. (2018) "The Role of Polygamy in Social Stratification," (*Journal of Social Sciences in Africa*). 33

[40] Karen, Stacey. (2020)"Societal Expectations and Infertility," (*Journal of Social Issues* 76, no. 2 (2020): 405–415. Rebecca Jones. (2018) "Reproductive Responsibility: A Gendered Issue," (*Gender Studies Review* 15, no. 1 (2018): 45–60.

well-being.[41]In numerous communities, when issues of infertility emerge, the blame predominantly rests on women, causing substantial psychological and emotional distress. This prevailing societal tendency not only intensifies the challenges encountered by women but also perpetuates detrimental stereotypes concerning gender roles in reproductive health.[42]Such expectations foster a stigma surrounding infertility, frequently culminating in feelings of inadequacy and guilt. This stigma further prevents individuals from obtaining the necessary support during difficult times.

[43]Moreover, cultural norms and values are pivotal in family planning, particularly within rural communities. These elements frequently impact the decisions made by families concerning the timing and number of children they intend to have.[44]Within these communities, there tends to be a prevailing belief that larger families are associated with enhanced prosperity and labor resources for agricultural purposes. This conviction may shape family planning choices and influence communal values on support for larger households.[45]Consequently, such beliefs can dissuade the utilization of contraception, thereby reinforcing the perception that having numerous children is vital for economic stability and success.[46]This situation can impede women's reproductive autonomy and restrict their access to family planning resources, perpetuating a cycle of reliance on traditional family structures. These limitations impact individual decisions and resonate on a broader societal scale, reinforcing outdated norms related to gender roles and tend not to focus on the rigours the children will face.

---

[41] Jennifer, Smith. (2019). "The Stigma of Infertility," (*Health & Ethics* 17, no. 1 (2019): 88–100, Centre for Bioethics Research).

[42] Melissa, Brown. (2021) "The Psychological Effects of Infertility," (*Journal of Women's Health* 30, no. 4 (2021): 321–330.

[43] Harriet, Williams. (2022). "Cultural Norms and Family Planning," (*Journal of Rural Studies* 54 (2022): 150–160.

[44] Laura, Davis. (2020). "Cultural Perspectives on Family Size,"(*Agriculture and Society* 12, no. 2 (2020): 214–225.

[45] Sophia, Taylor. "Family Planning Practices in Rural Areas," (*Rural Sociology* 82), no. 3 (2017): 450–463.

[46] Amanda, Miller. (2021). "Reproductive Rights and Family Planning," *Women's Studies Quarterly* 39, no. 3 (2021): 98–112.

6.10 Modern Influences and Changing Attitudes toward Procreation

[47]Urbanization, education, and various economic challenges have significantly altered procreation-related cultural expectations in contemporary society. These shifts underscore the need to reassess traditional family structures and roles in light of shifting societal norms.[48]As urban areas expand and access to education improves, more couples increasingly embrace family planning methods, shifting societal perspectives on childbearing. This evolution highlights the importance of adapting our social policies to support diverse family configurations and prioritize individual choices.[49]Furthermore, women are gaining greater autonomy in making reproductive decisions, empowered by their educational achievements and changing societal views, recognising their rights and choices. This empowerment allows women to make choices that align with their aspirations rather than conforming to traditional expectations.

[50]However, it is important to note that traditional norms and values profoundly influence many areas of life, creating a complex landscape where modern practices and longstanding beliefs sometimes coexist and clash. The tension between these evolving practices and traditional norms can result in societal debates and conflicts, particularly within various cultural and community contexts. Amid all the disasters and influenced by the aforementioned factors, population growth has been significant in the country, raising both opportunities and challenges for policymakers and communities alike, who must navigate the implications of this demographic shift in a rapidly changing world.

---

[47] Rebecca, Johnson. (2019). "The Impact of Education on Family Dynamics," (*Education and Society Journal* 15, no. 4 (2019): 267–299.

[48] Jane, Doe. (2021). "Family Planning in Urban Settings" (*Urban Studies Journal* 27), no. 3 (2021): 112–130.

[49] Julia, Adams. (2023). "Women and Reproductive Autonomy: An Exploration of New Dynamics" (*Journal of Social Issues* 29, no. 2 (2023): 123–145.

[50] Melissa, Clark. (2020). "Tradition vs. Modernity in Family Planning," (*Sociology Today* 38), no. 1 (2020): 46

6.11 Population Growth and Environmental Sustainability

[51]The findings from the 2022 Zimbabwe census reveal a significant demographic shift.[52]The population has increased by 16.2% over the previous decade, reaching approximately 15.1 million as of April 20, 2022.[53]This increase translates to an annual growth rate of 1.5%, which indicates a steady rise in the number of inhabitants. However, climatic disasters may jeopardize this growth, leading to displacement and negatively impacting population stability.[54]However, such growth presents numerous challenges, particularly in terms of resource management and environmental sustainability. This growth raises important questions about reproduction amid climatic disasters, which pose significant challenges to sustainable living and resource management. It urges us to reflect on the implications of increasing population numbers in an era of environmental uncertainty.

Due to the increasing number of people worldwide, it is essential to consider the implications on resources. The escalating population heightens the demand for essential natural resources, including water, land, and energy.[55]Environmental pressure grows as these resources become scarcer, resulting in possible degradation and negative ecological consequences. This highlights the urgent need for sustainable management practices to mitigate these negative impacts.[56]Moreover, this strain complicates efforts to achieve sustainability, as policymakers and citizenry must navigate the delicate balance between meeting current needs and preserving the environment for future generations. This challenge highlights the urgent need for innovative solutions that balance economic growth with environmental stewardship.

Considering these developments, it is imperative to consider the ethical implications associated with continued population growth, particularly as the planet grapples with the urgent realities of climate change. This situation calls for the re-evaluation of procreation practices, emphasizing the need to incorporate environmental stewardship into personal and societal decision-making. Given the interconnectedness of population growth and environmental health, promoting responsible family planning and raising awareness about the impact of reproductive choices on the planet is crucial to fostering a sustainable future.

---

[51] Zimbabwe National Statistics Agency. *Zimbabwe Census 2022: Population and Housing Census Results*. Available at: https://www.zimstat.co.zw (accessed February 2, 2025).

[52][53] Zimbabwe National Statistics Agency. *Zimbabwe Census 2022: Population and Housing Census Results*. Available at: https://www.zimstat.co.zw (accessed February 2, 2025).

[53] World Bank. *Population Growth (Annual %)*. Available at: https://data.worldbank.org (accessed February 26, 2025). 22

[54] United Nations Development Programme. *Sustainable Development Goals*. https://www.undp.org/sustainable-development-goals (accessed January 13, 2025).

[55] Rebecca, Johnson and Thomas, Lee. "Ecological Consequences of Resource Scarcity." (*Journal of Environmental Science* 15, no. 4 (2021)). 45.

[56] Angela, Davis. (2019). "*Sustainability and Resource Management*." (Environmental Studies Press, 2019).

6.12 African Communitarianism and Ethical Responsibility

[57]African communitarianism, a philosophical framework rooted in African traditions, emphasizes the importance of collective well-being over individual interests, making it a valuable lens for evaluating procreation ethics in climate change. This philosophy encourages a community-oriented perspective where decisions are made with consideration for how they impact the broader social and environmental fabric, particularly in terms of intergenerational responsibility.

[58]In the context of Zimbabwe, the principle of ubuntu acts as a fundamental ethical guide, emphasizing the moral obligation to consider not only the present population but also the well-being of future generations before making decisions that may have potential environmental consequences. Ubuntu, which translates to "I am because we are," reinforces the interconnectedness of all individuals and the responsibility that comes with it. As climate change increasingly threatens ecological integrity, the depletion of natural resources raises critical questions about the sustainability of procreation as an ethical act.

[59]The demand for resources grows with each new birth, compounding pressures on an already fragile environment.[60]Current environmental challenges, including deforestation, water scarcity, and biodiversity loss, emphasize the need to reconsider procreation in ethical discourse. They prompt us to reflect on how population growth contributes to environmental degradation and whether individuals should consider the ecological legacy they leave behind when deciding to have children.

As such, integrating African communitarianism and the principle of ubuntu into discussions about procreation and ethics can lead to a more holistic understanding of our responsibilities toward both the Earth and future generations. This perspective may promote a more sustainable ethos that balances human needs with ecological preservation. Therefore, we must continue exploring these ethical frameworks as we navigate the complexities of climate change and its implications for societal structures.

---

[57] Thaddeus, Metz. (2011). "Ubuntu as a Moral Theory and Human Rights in South Africa," (*African Human Rights Law Journal* 11, no. 2 (2011). 53–73.

[58] Musa, Zubairu. (2010). "Ubuntu and the Governance in Africa: Linking Tradition and Modernity," (*African Journal of Political Science* 5, no. 1 (2010)). 23.

[59] Joseph, Ndlovu. (2021). "Revisiting Procreation Ethics in an Era of Climate Change," (*Zimbabwe Journal of Sociology* 10, no. 2 (2021): 102.

[60] United Nations, *World Population Prospects 2019*, https://www.un.org/development/desa/pd/content/world-population-prospects-2019 (accessed March 12, 2025).

6.13 Intergenerational Obligations and Environmental Ethics

In the context of Zimbabwe, the stakes associated with intergenerational obligations are exceptionally high, as marginalized communities disproportionately bear the brunt of climate change's adverse effects.[61]According to Maposa, the current generation's responsibilities extend beyond mere survival; they encompass a proactive duty to institute sustainable practices to protect the environment and ensure a viable future for their communities. This includes promoting renewable energy sources, water conservation measures, and sustainable agricultural practices, vital in fostering resilience against climate change. Furthermore, the necessity of engaging marginalized voices in policy-making processes cannot be overstated, as their experience and insights are essential in crafting effective and equitable solutions.

The importance of sustainable development in addressing intergenerational justice cannot be over-emphasized. For instance, the United Nations Sustainable Development Goals (2015) advocate for the responsibility of current generations to leave a better world for future inhabitants. Similarly, research by Oxfam (2018) highlights the interconnection between environmental degradation and social justice, illustrating the need for comprehensive strategies that prioritize the rights and needs of vulnerable populations. These insights underscore the urgency of ethical considerations in ongoing climate change and procreation debates, particularly in contexts like Zimbabwe, where the impacts are most profoundly felt.

6.14 Moral Dilemmas in Procreation and Climate Change

[62]A fundamental ethical dilemma arises when considering whether procreation is justifiable in a world facing ecological crises. Philosophers such as Carter contend that bringing children into a deteriorating environment raises moral concerns regarding their quality of life and access to resources.[63]Conversely, African communitarian ethics suggest that procreation should be viewed not solely through an individualistic lens but rather as a communal responsibility where society collectively ensures the survival and well-being of its members. This perspective emphasizes the interconnectedness of individuals within the community, highlighting that the well-being of one is intrinsically linked to the well-being of all.

[61] Memory, Maposa. (2020). " *Intergenerational Ethics and Climate Justice in Africa*". (Cape Town: University of Cape Town Press). 119

[62] James, Carter. (2020). "*Ethics and the Environment: A Moral Inquiry*". (Oxford: Oxford University Press). 34

[63] Patrick Mugabe, "Sustainability and African Communitarianism: A Philosophical Inquiry," *Zimbabwe Journal of Philosophy* 5, no. 1 (2020): 51.

[64]While some advocate for population control measures to mitigate climate change, critics argue that such policies often overlook sociocultural contexts and the significance of lineage continuity in African societies. Therefore, an ethical balance must be struck between environmental sustainability and cultural values surrounding family and procreation.

6.15 Communal Approaches to Environmental Resilience

Rather than framing procreation as an ethical burden, African communitarianism offers a constructive approach by advocating for collective efforts in climate adaptation and environmental stewardship. For instance, various studies highlight the significanceofcommunityengagementintacklingclimate-relatedchallenges.[65]Strategies such as community-led afforestation, which help combat deforestation and create local jobs, can significantly contribute to sustainable development. Sustainable farming practices, particularly those prioritizing organic techniques, are vital for enhancing food security and maintaining ecological balance.[66]By engaging in these practices, communities can bolster their resilience against the adverse impacts of climate change. This commitment to environmental education empowers students with the knowledge needed to address ecological challenges. It instils a sense of responsibility toward the planet, cultivating informed and active citizens for the future.

[67]By leveraging traditional knowledge and communal solidarity, Zimbabwean communities can significantly contribute to alleviating the negative impacts of climate change without relying on Western population control measures, thus fostering a comprehensive approach to sustainability. This approach empowers local communities and promotes adaptive strategies that enhance resilience against environmental changes.

---

[64] Tafadzwa, Chigodora. (2019). 'Climate change and Rural Livelihoods In Zimbabwe: A socioeconomic Analysis'. (Journal of African Studies, 12(2)). 91

[65] Memory, Maposa. (2020). "*Intergenerational Ethics and Climate Justice in Africa*". (Cape Town: University of Cape Town Press). 119

[66] Simbarashe, Ndlovu. (2021). "Population Control and Climate Change: A Complex Relationship," (*Zimbabwean Journal of Sustainable Development* 12, no. 2 (2021): 108. Danai Chikodzi, "Sustainable Agriculture Practices in Zimbabwe," *Journal of Environmental Studies* 15, no. 3 (2022): 135.

[67] Simbarashe, Ndlovu. (2021). "Population Control and Climate Change: A Complex Relationship," (*Zimbabwean Journal of Sustainable Development* 12, no. 2 (2021): 108.

6.16 Conclusion

The ethics of procreation in climate change raise complex moral questions. In Zimbabwe, the link between environmental sustainability and cultural heritage is profound. Traditional values emphasize collective responsibility toward the environment, making this discourse relevant.

One practical framework is African communitarianism, which prioritizes community over individualism and stresses shared values and responsibilities. It encourages reflection on the implications of procreation in a climate-challenged world.

African communitarianism also highlights intergenerational obligations, urging current generations to ensure a liveable planet for future ones. It promotes communal resilience, calling on Zimbabwean societies to adopt sustainable practices that benefit the environment and community bonds.

By engaging in sustainable practices and collective environmental responsibility, Zimbabwean communities can tackle the moral dilemmas of climate change. These efforts address immediate concerns while preserving Zimbabwe's rich cultural identity. Balancing cultural traditions with ecological sustainability is vital for ensuring a flourishing future. Harmonizing procreation with environmental stewardship reflects respect for cultural values and global responsibility.

References

Adams, A. 2023a. Empowerment and reproductive choices. *Journal of Social Issues* 45 (2): 123–135.

Adams, J. 2023b. Women and reproductive autonomy: An exploration of new dynamics. *Journal of Social Issues* 29 (2): 123–145.

Adesina, T. 2014. Witchcraft, curses, and fertility: A cultural assessment. *Journal of Anthropology and Culture.*

Behrens, K. 2012. An African relational environmentalism and moral considerability. *Environmental Ethics* 34 (1): 75–93.

Brown, M. 2021. The psychological effects of infertility. *Journal of Women's Health* 30 (4): 321–330.

Carter, J. 2018. *Ethics and the environment: A moral inquiry*. Oxford: Oxford University Press.

Chakaipa, A. 2017. Cultural perspectives on family structures in Zimbabwe. *Journal of African Studies.*

Chawira, M. 2020. *Cultural values and practices in Zimbabwean communities*. Harare: ZIM Publications.

Chigodora, T. 2019a. Climate change and rural livelihoods in Zimbabwe: A socioeconomic analysis. *Journal of African Studies* 12 (2): 85–94.

Chigodora, T. 2019b. *The socioeconomic effects of climate change in Zimbabwe*. Bulawayo: National Book Publishers.

Chikodzi, D. 2022. Sustainable agriculture practices in Zimbabwe. *Journal of Environmental Studies* 15 (3): 45–67.

Chireshe, R. 2011. *Cultural practices influencing women's roles in Zimbabwean society*. Harare: Zimbabwe University Press.

Clark, M. 2020a. Tradition vs. modernity in family planning. *Sociology Today* 38 (1): 45–60.

Clark, T. 2020b. Tradition vs. modernity: Cultural expectations in family planning. *Cultural Studies Review* 24 (1): 45–67.
Davis, A. 2019. *Sustainability and resource management*. Environmental Studies Press.
Davis, L. 2020. Cultural perspectives on family size. *Agriculture and Society* 12 (2): 214–225.
Doe, A. 2021a. Family planning in urban contexts: New strategies for success. *Urban Studies Journal* 36 (3): 789–803.
Doe, J. 2021b. Family planning in urban settings. *Urban Studies Journal* 27 (3): 112–130.
Dube, D., and M. Chimonyo. 2018. Gender roles and childbearing in Zimbabwe. *Journal of African Studies* 45 (2): 134–150.
Dube, J. (2023). *Personal interview.*
Dube, T. 2021. The role of education in climate change awareness. *African Journal of Social Sciences* 10 (4): 34–50.
Elijah, A. 2019. Intersecting paths: Spirituality and medicine in African communities. *Journal of Health and Spirituality.*
FAO. 2022. *The state of food security and nutrition in the world 2022*. Food and agriculture Organization.
Johnson, L. 2019a. Education and women's autonomy. *Women's Studies Quarterly* 40 (4): 212–225.
Johnson, R. 2019b. The impact of education on family dynamics. *Education and Society Journal* 15 (4): 267–299.
Johnson, R., and T. Lee. 2021. Ecological consequences of resource scarcity. *Journal of Environmental Science* 15 (4): 234–245.
Jones, A. 2019. Cultural practices in family dynamics: The role of levirate marriage. *Journal of Ethnographic Studies* 12 (3): 45–58.
Jones, L. (2010). **Infertility as a social issue in African communities**. Journal of Social Issues.
Jones, R. 2018. Reproductive responsibility: A gendered issue. *Gender Studies Review* 15 (1): 45–60.
Lee, C. 2022a. Understanding family dynamics in the 21st century. *Journal of Family Research* 56 (2): 78–94.
Lee, K. 2022b. Personal choice and responsibility in family structures. *Journal of Family Theory & Review* 14 (2): 109–128.
Maposa, M. 2020a. *Intergenerational ethics and climate justice in Africa*. Cape Town: University of Cape Town Press.
Maposa, T. 2020b. Communitarian values and climate change: A Zimbabwean perspective. *International Journal of Climate Policy* 5 (2): 121–135.
Matlosa, K. 2015. Marriage, family structure, and social expectations in Zimbabwe. *African Studies Review* 58 (2): 33–54.
Mavengere, G. 2018. Polygamy and its impact on society. *Zimbabwean Journal of Sociology*.
Mavhunga, C. 2019. The sacred duty of procreation: Family and cultural preservation. *Zimbabwean Journal of Social Sciences* 27 (1): 58–75.
Metz, T. 2011a. *African philosophy: Theoretical perspectives*. New York: Oxford University Press.
Metz, T. 2011b. Ubuntu as a moral theory and human rights in South Africa. *African Human Rights Law Journal* 11 (2): 53–73.
Miller, A. 2021. Reproductive rights and family planning. *Women's Studies Quarterly* 39 (3): 98–112.
Mlambo, A. (2014). Gender roles and marriage in Zimbabwe. Bulawayo: Books of Zimbabwe.
Moyo, R. 2012. The spiritual dimensions of fertility in traditional settings. *African Journal of Spirituality.*
Moyo, R. 2021. Community resilience and adaptation strategies. *African Journal of Environmental Management* 8 (1): 88–102.
Muchemwa, S. 2023. *Personal interview*.
Mudavanhu, B. 2018. The role of polygamy in social stratification. *Journal of Social Sciences in Africa*.

Mudzamiri, V. 2020. Socioeconomic factors influencing polygamous marriages. *African journal of family Studies.*
Mugabe, P. 2020a. Sustainability and African communitarianism: A philosophical inquiry. *Zimbabwe Journal of Philosophy* 5 (1): 44–56.
Mugabe, R. (2020b). Climate change and its impact on Zimbabwean agriculture. Harare: Zimbabwe University Press.
Mujere, N. 2021. Economics of family legacy in Zimbabwe. *Journal of Economic and Social Research* 33 (3): 200–215.
Munsaka, T. 2021. Child mortality and family dynamics in Zimbabwe. *International journal of health Studies.*
Ndlovu, J. 2021a. Revisiting procreation ethics in an era of climate change. *Zimbabwe Journal of Sociology* 10 (2): 95–112.
Ndlovu, R. 2019. Agricultural practices and family labor in Zimbabwe. *Zimbabwean Agricultural Review.*
Ndlovu, S. 2010. Lobola: The bride price as an economic and cultural practice. *Journal of Southern African Studies* 36 (3): 479–494.
Ndlovu, S. 2021b. Ethics, climate change, and the future of Zimbabwe. *African Journal of Ethics* 8 (1): 100–112.
Ndlovu, S. 2021c. Population control and climate change: A complex relationship. *Zimbabwean Journal of Sustainable Development* 12 (2): 103–115.
Nichols, P. 2018. Beliefs and practices surrounding infertility. *Journal of Cross-Cultural Research.*
Ogunyemi, O. 2007. African cultures and fertility. *Journal of Cultural Perspectives.*
Okyere-Manu, B. 2015. Cohabitation in Akan culture of Ghana: An ethical challenge to gatekeepers of the indigenous knowledge system in the Akan culture. *Alternation Special Edition* 14:45–60.
Sibanda, L. 2022. *Gender expectations and family dynamics in Zimbabwe*. University of Zimbabwe Press.
Smith, J. 2015. Familial honour and childbearing: A cultural analysis. *Sociology of Family Studies.*
Smith, J. 2019. The stigma of infertility. *Health & Ethics* 17 (1): 88–100.
Smith, J. 2020a. Population growth and resource consumption. *Global Journal of Resources* 8 (2): 56–67.
Smith, L. 2020b. Urbanisation and its effect on cultural values. *Urban Sociology* 12 (2): 89–102.
Smith, R. 2020c. Cultural expectations in modern society. *Culture and Society Review* 12 (1): 1–15.
Smith, R. 2020d. Tradition and conception: The influence of family in fertility issues. *International Journal of Sociology* 45 (2): 101–113.
Stacey, K. 2020. Societal expectations and infertility. *Journal of Social Issues* 76 (2): 405–415.
Taylor, K. 2016. Seeking help: A study of medical and spiritual practices. *International Journal of Health Studies.*
Taylor, M. 2021. The interplay of family and fertility: Alternative solutions to infertility. *Family Relations* 70 (4): 599–612.
Taylor, S. 2017. Family planning practices in rural areas. *Rural Sociology* 82 (3): 450–463.
UNICEF. 2022. *Education disrupted: The impact of emergencies on children's learning*. United Nations Children's Fund. https://www.unicef.org.
United Nations (2019). *World population prospects 2019*. Available at: https://www.un.org/development/desa/pd/content/world-population-prospects-2019 Accessed: 12 March 2025.
United Nations Development Programme (2022). *Sustainable development goals*. Available at: https://www.undp.org/sustainable-development-goals Accessed: 13 January 2025.
WHO. (2023). *Malnutrition in children: Key facts and global response*. World health Organization https://www.who.int
Williams, H. 2022. Cultural norms and family planning. *Journal of Rural Studies* 54:150–160.
World Bank (2022a). *Population growth (annual %)*. Available at: https://data.worldbank.org Accessed: 26 February 2025.

World Bank. 2021. *Climate change and agriculture: Impacts on food security in developing countries*. World Bank Publications.

World Bank. (2022b). *World development indicators: Population growth and demographics*. https://data.worldbank.org.

Zimbabwe National Statistics Agency (2022). *Zimbabwe Census 2022: Population and Housing Census Results*. Available at: https://www.zimstat.co.zw Accessed: 2 February 2025.

Zimbabwe Red Cross. (2023). *Statement on flood relief efforts*.

Zubairu, M. 2010. Ubuntu and the governance in Africa: Linking tradition and modernity. *African Journal of Political Science* 5 (1): 23–34.

Part III
Challenges and Implications

Chapter 7
Necroeconomic Population Policy, the Racialized Capitalist World and the Surplus Population of Disposable Beings

Peter Mwipikeni

Abstract Recent scholarly work has established that a reconfiguration of population control occurred following the 1994 International Conference on Population and Development in Cairo. The Cairo conference was hailed as representing a discursive paradigm shift from the control of population growth by targeting the reduction of fertility rates of women in the Global South to women's empowerment associated with the promotion of women's reproductive rights. Some scholars hold that the discursive shift did not lead to the demise of population control, but rather to the reconfiguration of multiple forms of population control, including geopopulationism, biopopulationism, and demopopulationism. In addition, some scholars show the prevalence of necropopulationist control which is associated with reproductive technologies and agricultural technologies that lead to harmful effects on health, slow death, and the toxification of landscapes in the Global South. In this chapter, I will show that necroeconomic populationism is, in fact, another ongoing form of racialized population control which promotes the welfare of the population of the Global North while undermining the welfare of the population of Africa. I will show that ongoing forms of racialized population control are inevitable and enduring outcomes of a racialized world order that is configured by coloniality of being that others Africans and other non-Western people as surplus disposable beings.

Keywords Population control · Necroeconomic population policy · Surplus population of disposable beings · Racialized world order · Capitalism

P. Mwipikeni (✉)
Department of Philosophy and Ethics Studies, University of KwaZulu-Natal, Pietermaritzburg, KwaZulu-Natal, South Africa
e-mail: MwipikeniP@ukzn.ac.za

B. Okyere-Manu, P. Mwipikeni (eds.), *An African Approach to Population Ethics*, Advancing Global Bioethics 23,
https://doi.org/10.1007/978-3-032-24464-2_7

7.1 Introduction

There are two types of population control policies: those that seek to tackle overpopulation by seeking to reduce the fertility rates of poor, marginalized women in the Global South;[1] and those that promote the welfare of the population of the Global North by reducing the welfare of the population in Africa and other parts of the Global South. In 1994, Cairo hosted the International Conference on Population and Development, a gathering which was hailed as representing a discursive paradigm change from the lexicon of fertility control to a new grammar of the promotion of women's reproductive rights and health.[2] The policies of population control that sought to reduce the fertility of women in the Global South were discredited as an unjust violation of human rights.

However, at the Cairo Conference, there was a focus on the discrediting of coercive population control measures that seek to reduce the fertility rates of women in the Global South. Population control measures that seek to promote the welfare of the population in the Global North while undermining the welfare of the population in the Global South were not placed on the agenda. Bhatia et al. hold that the Cairo Conference of 1994 did not lead to the demise of population control.[3] There was an erasure of the language of population control, without an ending of the activities associated with population control.[4]

For Bhatia et al., the reconfiguration of population control after the Cairo conference is associated with a series of related terms: "biopopulationism," "geopopulationism," and "demopopulationism."[5] Shaw and Wilson hold that "necropopulationism" is another ongoing form of population control.[6] "Biopopulationism" is associated with seeking a high quality of life, which, among other things, is associated primarily with bearing children judiciously and using selective reproduction technologies to birth high-quality children.[7] Biopopulationism is also associated with the promotion of the adoption of an anti-natalist bias by poor, non-white women in the Global South. Such an anti-natalist bias equates "responsible, risk-avoiding and empowering behaviour with postponing, reducing, spacing and timing births."[8] By contrast, middle-class white women are

[1] Bhatia, R. Sasser, S. J. Ojeda, D. Hendrixson, A. Nadimpally, S. and Foley, E. E. 2019. A feminist exploration of 'populationism': engaging contemporary forms of population control. *Gender, Place & Culture* 27: 333–350.

[2] Richey, L. A. 2010. Reproductive Health, Family Planning, and HIV/AIDS: Dangers of (Dis) Integration in Tanzania and Uganda In *Markets and Malthus Population, Gender, and Health in Neo-liberal Times*. Eds Mohan Rao and Sarah Sexton, 265–298. New Delhi: Sage Publications.

[3] Bhatia et al. A feminist exploration of 'populationism'.

[4] Bhatia et al. A feminist exploration of 'populationism'.

[5] Bhatia et al. A feminist exploration of 'populationism'. 1.

[6] Shaw, A. and Wilson, K. 2020. The Bill and Melinda Gates Foundation and the necro-populationism of 'climate-smart' agriculture. *Gender, Place & Culture* 27: 370–393. 370–371

[7] Bhatia et al. A feminist exploration of 'populationism'. 9–10.

[8] Bhatia et al. A feminist exploration of 'populationism'.11.

increasingly encouraged to have children.[9] A "demopopulationist" policy is associated with "population engineering" that seeks the alteration of demography in a way that leads to the appropriate population size and composition.[10] Among the aspects of demopopulationism are the suppression of the fertility of poor women in the Global South and promotion of the fertility of middle-class women, and the promotion of the constitution of self-regulating subjects who control their fertility to achieve desirable social and economic ends.[11] Another feature of demopopulationism is the production of racialized statistical data which identify women in the Global South as having problematic bodies that should be targeted for fertility control.[12] Geopopulationism is associated with the management of space and control of mobility.[13] Bhatia et al. hold that geopopulationist measures are historically associated with colonial conquest, colonialism, racism, and the "uneven geography of resource allocation."[14] Necropopulationist policies undermine life by making it vulnerable to destruction by the promotion of "green revolution" technologies that emit endocrine chemical substances, harmful reproductive technologies, displacements, indebtedness, and toxification that leads to "dispossession, biological disruption, and death."[15] One of the examples of necropopulationist policies is the death of women in sterilization camps in India.[16]

Bhatia et al. reflect that population control was not relegated to history after the Cairo Conference.[17] Indeed, they show promotion of various forms of population control after this event. There is, at present, a lack of adequate attention on reasons for the persistence of various forms of population control. Bhatia et al. only problematize "multiple processes which render humanity abstract and relative by placing it in opposition to nature or the economy."[18] If there is a disequilibrium between the environment and population growth, or between the number of human beings and the economy, then there are interventions that seek to reduce human numbers. Furthermore, Bhatia et al. deploy populationism to show multiple ways of "understanding and addressing social and environmental problems that are dis/similar to Malthusianism and prone to mechanisms that produce, reinforce and naturalize inequalities along lines of race, class, gender, and place."[19] However, it is unclear whether the ways of thinking that render humanity relative and abstract, as well as Malthusian thinking, can help in understanding the perpetuation of the promotion of

[9] Bhatia et al. A feminist exploration of 'populationism'.11.

[10] Bhatia et al. A feminist exploration of 'populationism'. 5.

[11] Bhatia et al. A feminist exploration of 'populationism'. 5.

[12] Bhatia et al. A feminist exploration of 'populationism'. 5.

[13] Bhatia et al. A feminist exploration of 'populationism'. 7.

[14] Bhatia et al. A feminist exploration of 'populationism'. 7.

[15] Shaw, A. and Wilson, K. The Bill and Melinda Gates Foundation, 381–382.

[16] Shaw, A. and Wilson, K. The Bill and Melinda Gates Foundation.

[17] Bhatia et al. A feminist exploration of 'populationism'.

[18] Bhatia et al. A feminist exploration of 'populationism'. 4.

[19] Bhatia et al. A feminist exploration of 'populationism'. 4.

population control after the Cairo Conference. This chapter seeks to contribute knowledge that deepens understanding of factors that sustain the persistence of population control measures after Cairo and ways of terminating these population control measures.

In this chapter, I will show that the persistence of population control policies results from the existence of a racialized capitalist world order.[20] Population control measures are impelled by a racialized capitalist world order that promotes the well-being of the Global North's population of human beings. Throughout history, concerns about population growth in the Global South was related to the preservation of Western capitalist accumulation practices that promote well-being in the Global North. The concern with overpopulation was not about the threat to the biosphere as a global common. The biosphere in a racialized world is a racialized entity that exists to support the population of human beings in the Global North. The biosphere took on a racialized form and was constituted as a primarily Global North property to promote the well-being of people in the Global North from the time when the racialized capitalist world came into being at the beginning of the modern era. The concern with overpopulation is an attempt to reduce population growth in the Global South, which leads to a substantial population of angry young men who can be recruited by terrorist groups that are a security threat to the welfare of the population of the Global North.[21] As further evidence, a document published by the US government, the "US National Security Study Memorandum 2000: Implications of Worldwide Population Growth for US Security and Overseas Interests," indicated that overpopulation is a threat to Western accumulation practices, as poor people in the Global South are a threat to the US corporate economic interests in the following ways.[22] The many young people in the high-fertility populations of the Global South may direct their attacks at foreign multinational corporations.[23] Additionally, some countries in the Global South may advocate for fair distribution of wealth.[24] There may also be threats to the concessions granted to foreign multinational corporations

---

[20] Bhatia et al. hold that population control was not relegated to history. Bhatia et al. prefer the use of populationism to refer to various ongoing forms of population control. They seek to conform with the putative discursive paradigm shift from population control to reproductive rights and health. Such conformism is problematic when the language of population control is erased, despite the ongoing activities of population control. The Family Planning 2020 (FP2020) initiative targeted enrolling "120 million new contraceptive users by the year 2020" (Bhatia et al. 2019, 6). For Bhatia et al., this produces a sense of urgency and incentivizes rapid scale-up of access to potentially harmful long-acting reversible contraceptives. Bhatia et al. disavow the use of the term of population control while at the same identifying various ongoing forms of population control. This amounts to a contradiction. In this chapter, I will use the term population control as population control was not terminated by the 1994 Cairo Conference. Bhatia et al. A feminist exploration of 'populationism'.

[21] Wilson, K. 2017. Re-centring 'race' in development: population policies and global capital accumulation in the era of the SDGs. *Globalizations* 14:432–449.

[22] Wilson, K. Re-centring 'race' in development. 438.

[23] Wilson, K. Re-centring 'race' in development.

[24] Wilson, K. Re-centring 'race' in development.

(Wilson 2017). The discourse on overpopulation is related to the fear that the population growth of poor people in the Global South can lead to quests to "deracialize" the Earth's biosphere. A racialized world distributes benefits to the Global North and burdens to the Global South, which it was specifically designed to do. The main/primary feature of population control that emerges in a racialized world is associated with promoting the well-being in the Global North. Population control policies are harmful as they are associated with promoting the welfare of the population in the Global North while simultaneously undermining the welfare of the population in the Global South. Additionally, population control is linked to demographization, which attributes poverty to high birth rates, which lead to overpopulation.[25] This conceals the complicity of global capitalist accumulative practices that cause poverty and undermine the welfare of the population by looting wealth from the extractive sector in African countries and other parts of the Global South.

In what follows I will show that population control measures are structural effects of a racialized capitalist world. I will show the role of capital and race in the reproduction and maintenance of population control measures. Also, I will show another ongoing form of population control, which I refer to as necroeconomic populationism, that has not received attention. I will show that the primary feature of necroeconomic population control measures in a racialized world order pertains to controlling the world economy (natural resources and wealth) so as to promote the welfare of the population in the Global North by undermining the welfare of the population in Africa and the other parts of the Global South.

I will show how necroeconomic population policy is used to constitute economies in Africa in a way that undermines the welfare of the African population and promotes the welfare of the Global North populations.[26] In the next section, I will show that the persistence of racialized population control policies results from a racialized world order. In the final section of the chapter, I will demonstrate that the termination of population control policies is contingent on configuring a deracialized and non-capitalist world order.

---

[25] Schultz, S. Demographic futurity: How statistical assumption politics shape immigration policy rationales in Germany. https://www.academia.edu/36758617/Demographic_futurity_How_statistical_assumption_politics_shape_immigration_policy_rationales_in_Germany_2018_

[26] I acknowledge that there is also a transfer of wealth from Africa to the other parts of the Global South such as China, India, Brazil and South Africa that enrichens elites from these countries and a large section of the population in China. However, such transfer is not part of the scope of this chapter. I will focus on the transfer of wealth mainly from Africa to the Global North.

7.2 Necroeconomic Population Control Policy

Bhatia et al. hold that population control is not only about fertility control. It is also about the restriction on "family and community formation and survival" and welfare restriction, among other phenomena.[27] In what follows I will use the term "population control" in its broad sense. I will focus on necroeconomic population control, which involves configuring economies in the Global South to promote the welfare of the population in the Global North while simultaneously undermining the welfare of the population in the Global South. The concept of necroeconomic population policy highlights the destructive aspects of global population control policies that undermine the well-being of a population in one region of the world to promote the welfare of the population in another region. The destructive aspects involve the historical contingent construction of a racialized population of disposable beings. Here I utilize Mbembe's conception of necropolitics and Shaw and Wilson's conception of necropopulism to conceptualize the necroeconomic population policy. I will briefly explain Mbembe's conception of necropolitics and Shaw and Wilson's conception of necropopulationism to clarify the notion of necroeconomic populationism.

7.3 Mbembe's Concept of Necropolitics

For Mbembe, necropolitics refers to the use of power to deploy processes and weapons that destroy human bodies and populations.[28] This leads to the creation of racialized sites of death involving the deployment of racialized and colonial violence.[29] Another idea that Mbembe connects to necropolitics pertains to the constitution of "death worlds" which are characterized by social conditions that confer on populations the status of the "living the dead" (Mbembe 2019, 92). In his analysis, Mbembe links necropolitics with biopower. Biopower involves the politics of death, which involved dividing the population into groups that should live and those that are deemed superfluous and disposable. Such splitting of the population by biopower is structured along racial lines. The significant role of race in the "economy of biopower" involves permitting the destruction of human lives and the regulation of the distribution of death (Mbembe 2019, 71).

Shaw and Wilson appropriate Mbembe's conception of necropolitics and deploy it to construe necropopulationism. Necropopulationism is associated with processes and technologies that target racialized and gendered populations in the Global South

[27] Bhatia et al. A feminist exploration of 'populationism'. 3.

[28] Mbembe, A. 2019. *Necropolitics*. Translated by Steven Corcoran. Durham and London: Duke University Press. 68, 92.

[29] Shaw, A. and Wilson, K. The Bill and Melinda Gates Foundation, 382.

with dispossession, "toxification," and death.[30] Among the examples that show how necropopulationism functions include the women who died in sterilization camps in India, and also the "toxification of landscapes" by donated agricultural technologies such as genetically modified seeds and fertilizers, which cause prove harmful to health.[31]

Mbembe's necropolitics focuses on the deployment of power to promote the life and welfare of others, while undermining the life and welfare of disposable beings in the Global South. Similarly, Shaw and Wilson's necropopulationism focuses on the use of power to dispossess people in the Global South and the use of harmful reproductive and agricultural technologies that harm the lives of poor people in the Global South. Here I will draw on both Mbembe's conception of necropolitics and Shaw and Wilson's conception of necropopulationism to develop a concept of necroeconomic populationism. Necroeconomic populationism focuses on the configuration of entire economies in Africa and other parts of the Global South to promote the welfare of the population in the Global North and to undermine the welfare of the population in the Global South. I will utilize the concept of necroeconomic population policy to deepen Mbembe's necropolitics and Shaw and Wilson's necropopulationism by demonstrating how economies in the Global South are destructively constituted to undermine welfare and produce a population of disposable beings in the Global South, while simultaneously creating a population of indispensable and valuable beings in the Global North. Now, I will focus on how necroeconomic population policy is used to configure economies in Africa to promote the welfare of the population in the Global North.

The necroeconomic population policy functions as a racialized population constitutive mechanism for forming the surplus population of disposable beings in Africa and other parts of the Global South and a population of indispensable and valued human beings in the Global North. Necroeconomic population policies reduce the welfare of the population in Africa to promote the welfare of the populations in the Global North. This creates a population of disposable waste. This is a population that can be disposed of as waste and which is treated as superfluous in the promotion of the welfare of the population of the Global North. The recipients of the benefits that emanate from undermining the welfare of the population in Africa and other parts of the Global South emerge as a population of indispensable and valued human beings. One can refer in this context to the necroeconomic measures that differentiate, sustain, and reproduce the two population groups as population control policies.

Necroeconomic population measures do not seek to control natural births by limiting the fertility rates of poor women in Africa. Rather, necroeconomic population policy influences the historical and economic structural conditions that shape the creation of two distinct population groups. The necroeconomic population policies take a geospatial structural form associated with shaping natural procreation to

[30] Shaw, A. and Wilson, K. The Bill and Melinda Gates Foundation, 372.

[31] Shaw, A. and Wilson, K. The Bill and Melinda Gates Foundation, 381–382.

produce two different types of population groups in different world zones. The necroeconomic policy ensures that those who procreate in Africa produce human beings who are likely to belong to the population group of superfluous disposable beings, and that those who procreate in the Global North produce human beings who belong to the population group of indispensable valuable beings. The necroeconomic population control structurally configures the world to ensure that procreation in Africa produces a population of disposable beings and procreation in the Global North produces indispensable and valuable beings. Necroeconomic populationist measures control how human beings are produced in the Global South and the Global North..

The structural configuration of the world to control global outcomes of procreation involves the necroeconomic control of economies in Africa and other parts of the Global South to promote the well-being of populations in the Global North. African economies are controlled and used to promote the welfare of the population in the Global North. Such conditions structurally constitute procreation to produce disposable beings in Africa. The transfer of wealth essential for the welfare of the population in Africa and the other parts of the Global South to the Global North condemns procreation in some parts of the Global South in order to produce some disposable beings.

I will now show the necroeconomic ordering of the economies in Africa to promote the welfare of the population in the Global North. Europe experienced an overaccumulation crisis in the 1800s.[32] Such an overaccumulation crisis undermines the capitalist goals of businesses which involve maximizing profits. The solution involved what Patrick Bond refers to as a "geospatial fix" of the displacement of the crisis from Europe to Africa.[33] Such geographical displacement involved the colonial conquest of Africa to loot the wealth and plunder the resources of that continent.[34] Such a geospatial fix involved the necroeconomic ordering of the economy of Africa, as the looting of the wealth of Africa undermined the welfare of Africa and promoted the welfare of Europeans.

In addition, in the postcolonial era such necroeconomic ordering of the economies in Africa, which transfers the wealth of Africa to promote the welfare of the population of the Global North while at the same time undermining the welfare of the population in Africa, is associated with the foreign control of accumulation process in African countries by the multinational corporations from the Global North and, recently, from the emerging economies. One of the conditions necessary for development that can promote the welfare of the population in an African country is

[32] Bond, P. 2019. Luxemburg's Critique of Capital Accumulation, Reapplied in Africa. *Journal Für Entwicklungspolitik* XXXV:92–118.

[33] Bond, Luxemburg's Critique of Capital Accumulation.

[34] Bond, P. 2019. Luxemburg's Critique of Capital Accumulation.

control of the national accumulation process.[35] Developed countries that promote the welfare of their populations have control of the national accumulation process.[36].

Such control of the national accumulation process involves five distinct factors. The first of the factors is control of the social reproduction of labour.[37] The "reproduction of labour" involves a viable agricultural sector that produces surplus food.[38] The second factor is control of national surplus and capital, and the capacity of the nation-state to direct investment.[39] The other factors include control of access to natural resources, control of technology, control of the market, and competitiveness in the global market in at least one product.[40]

However, most African countries lack national sovereign control of their own natural resources.[41] It is foreign multinational corporations that dominate the extraction of natural resources in most African countries.[42] The foreign multinational corporations repatriate the profits.[43] Such repatriation is part of the transfer of wealth to the Global North, which promotes the welfare of the population in the Global North and undermines the welfare of the population in Africa.

The necroeconomic ordering of the African economies has a racialized form associated with the promotion of the welfare of the population in the Global North and, simultaneously, undermining the welfare of the population in Africa. The undermining of the welfare of Africa's population leads to the formation of a population of disposable beings. I will now focus on the creation of a surplus population of disposable beings by necroeconomic population policy. For Michelle Yates, such human beings in the surplus population do not have the means for subsistence and may survive by "trash pick" and on the "black market."[44] For Mike Davis,

This outcast proletariat—perhaps 1.5 billion people today, 2.5 billion by 2030—is the fastest growing and most novel social class on the planet. By and large, the urban informal working class is not a labor reserve army in the nineteenth-century sense: a backlog of strikebreakers during booms; to be expelled during busts; then reabsorbed again in the next expansion. On the contrary, this is a mass of humanity structurally and biologically redundant to global accumulation and the corporate matrix.[45]

---

[35] Amin, S. 1990. *Delinking: Towards a Polycentric World*, translated by Michael Wolfers. London: Zed Books.

[36] Amin, *Delinking: Towards a Polycentric World*.

[37] Amin, *Delinking: Towards a Polycentric World*.

[38] Amin, *Delinking: Towards a Polycentric World*. 10–11.

[39] Amin, *Delinking: Towards a Polycentric World*.

[40] Amin, *Delinking: Towards a Polycentric World*.

[41] Bond, Luxemburg's Critique of Capital Accumulation.

[42] Bond, Luxemburg's Critique of Capital Accumulation.

[43] Bond, Luxemburg's Critique of Capital Accumulation.

[44] Yates, M. 2011. The Human-As-Waste, the Labor Theory of Value and Disposability in Contemporary Capitalism. *Antipode* 43:1679–1695. 1680.

[45] Cited in Yates, The Human-As-Waste, 1679.

The necroeconomic population policy is associated with configuring African economies to promote the welfare of populations in the Global North by transferring wealth from Africa to the Global North. Such transference reduces the welfare of African populations and increases the welfare of populations in the Global North. The populations in Africa are rendered disposable as their welfare is undermined to promote the welfare of populations in the Global North.

Additionally, the necroeconomic population policy creates a surplus population of disposable beings in Africa by perpetuating an uneven global capitalist economy that allocates employment opportunities in a manner that benefits the Global North and undermines the welfare of African populations. The uneven development of the global capitalist system is associated with high rates of unemployment in Africa and other parts of the Global South, as well as low rates of unemployment in the Global North. According to the International Labour Organization Report, "in 2023, 53 million youth in sub-Saharan Africa were NEET (not in employment, education or training)."[46] In other words, "that means that more than 1 in 5 youth (21.9 percent) in the region were in NEET status."[47] The International Labour Organization states that the lack of productive and decent jobs remains the largest labour market challenge for young people in Sub-Saharan Africa. At 21.9% in 2023, the youth NEET (not in employment, education, or training) rate in sub-Saharan Africa exceeded the global rate (at 20.4%). The region is one of only three global regions judged to be 'off track' in its SDG commitment (under target 8.6) to lower numbers of young NEETs.[48]

Such an asymmetrical pattern of distribution of employment opportunities globally results from uneven trade and capital structures. World trade benefits the Global North. This generates substantial capital for reinvestment, which creates jobs in the Global North. There is a huge investment gap between the developed countries and the African countries.[49] According to the UNCTAD report, the developed countries are receiving about 10 times more direct investment per capita than the developing countries.[50] In 2021, Foreign Direct Investment (FDI) flows to developing economies grew more slowly than those in developed regions.[51] Africa's share of capital for industrialization is low compared to the share of the West. Additionally, African countries earn less from the trade sector of the world economy. There are also huge inflows of capital in the Global North and meager inflows of capital in the Global South. The earnings of Africa from the exports of primary products are not

[46] International Labour Report. 2024. Global Employment Trends for Youth 2024 Sub-Saharan Africa. https://www.ilo.org/sites/default/files/2024-08/Sub-Saharan%20Africa%20GET%20Youth%202024_0.pdf

[47] International Labour Report, Global Employment Trends.

[48] International Labour Report, Global Employment Trends.

[49] UNCTAD. *Key Statistics and Trends in Regional Trade in Africa*. Geneva: United Nations, 2019. V.

[50] UNCTAD. *Key Statistics and Trends*, 37.

[51] UNCTAD. *World Investment Report*. New York: United Nations Publications, 2022. X1.

sufficient for investment in job creation. This sustains an uneven distribution of employment opportunities globally. There are few employment opportunities in Africa.

Moreover, the transference of value from Africa to the Global North undermines capital formation in Africa for reinvestment in job creation. The transference of such value occurs when capitalist businesses adopt labour-saving technology. The adoption of labour-saving technology when there is the transfer of value from the Global South to the Global North creates high unemployment rates that amplify the production of disposable beings. According to Yates, labour is the source of value for the capitalist system.[52] The capitalist system extracts surplus value from labour.[53] Surplus value is the source of profits.[54] However, large-scale industries seek to increase the productivity of labour by introducing machinery.[55] This leads to a reduction in the number of employees.[56] Such a reduction creates a permanent surplus population that is excluded from the capitalist system of production, wages, markets, and exchange.[57] For Karl Marx, the capitalist law of accumulation links the increase in the accumulation of wealth by capitalists with the wasting of many human beings who are dumped into a rapidly growing population of permanently superfluous human beings.[58] Marx refers to the abovementioned process of the wasting of human beings as the "capitalist law of population." He states that "the working population therefore produces both accumulation of capital and the means by which it is itself made relatively superfluous, and it does this to the extent which is always increasing. This is the law of population peculiar to the capitalist mode of production."[59]

The production of a population of superfluous human beings by a capitalist system of production that replaces human jobs with automation is largely geographically confined to Africa and other parts of the Global South by necroeconomic policy that transfers wealth from Africa to the Global North. The transfer of wealth from Africa and other parts of the Global South to the Global North increases capital investments, which create jobs that compensate for the jobs eliminated by automation in the Global North. The creation of a surplus population of human beings who are excluded from the capitalist market and exchange by automation that takes human jobs is displaced to Africa. Africa and other parts of the Global South are the "dumping sites" of the crisis of the creation of a surplus population of disposable beings by the global capitalist system.

[52] Yates, The Human-As-Waste.

[53] Yates, The Human-As-Waste.

[54] Yates, The Human-As-Waste.

[55] Yates, The Human-As-Waste.

[56] Yates, The Human-As-Waste.

[57] Yates, The Human-As-Waste.

[58] Yates, The Human-As-Waste.

[59] Marx cited in Yates, The Human-As-Waste, 1689.

Additionally, the creation of a surplus population of disposable human beings in Africa is a result of the transfer of surplus value from Africa to the Global North. For Amin, such unequal transfer is not only a result of the exploitation of labour. It is also a result of the exploitation of natural resources.[60] The profits from the super-exploitation of labour and natural resources in Africa by foreign multinational corporations are transferred to the Global North.[61] There is a lack of reinvestment in the productive sector in Africa by foreign multinational corporations, which transfer surplus value to the Global North and the emerging economies. The lack of reinvestment in the productive sector in Africa undermines job creation, leaving many marooned in the surplus population of disposable beings.

There is a second scramble for African natural resources that involves the looting of African wealth. Such looting takes the form of green imperialism. Green imperialism entails the looting of wealth from the extraction of low-carbon metals, such as cobalt, manganese, and lithium, which are used to manufacture electric car batteries.[62] In the Democratic Republic of Congo (DRC), there is the looting of wealth from the extraction of cobalt by Chinese and Australian companies.[63] Such looting drains away capital that could have been used for job creation and the expansion of the productive sector. If such looting undermines reinvestment and job creation, then there is no job creation. The consequence is the expansion of a surplus population of human beings who cannot be exploited by capital and who are excluded from capitalist exchange, wages, and markets.

The uneven distribution of employment opportunities associated with the global capitalist system, which condemns African countries to high unemployment rates, creates a fertile ground for producing a population of superfluous and disposable beings. Superfluous and disposable beings cannot secure jobs and cannot access capital to engage in income-generating projects. Such human beings cannot own a house and cannot eat nutritious food. They live in squalid conditions and have no access to good-quality health care. Neoliberal austerity measures associated with underfunding health care, education, and utilities condemn the population of superfluous disposable beings to excessive suffering.

The necroeconomic population control measure is a population control measure in the sense that it reproduces many human beings in Africa and other parts of the Global South as disposable beings. It creates structural ordeals associated with exclusion from the capitalist market, exchange, and wages. Such structural socio-economic difficulties contribute to the disposal of many human beings into the population of superfluous beings. They may wield agency to oppose their construction as disposable beings. However, the structural configuration of the world by necroeconomic logic constitutes and reproduces human beings in Africa and other

---

[60] Amin, *Delinking: Towards a Polycentric World.*

[61] Bond, Luxemburg's Critique of Capital Accumulation.

[62] Radley, B. 2023. Green imperialism, sovereignty, and the quest for national development in the Congo. *Review of African Political Economy* 50(177–178):322–339.

[63] Radley, Green imperialism, sovereignty.

parts of the Global South as disposable beings. Necroeconomic population policies control the births of human beings into a population of disposable beings. They limit the number of human beings who can be born and integrated into the population of indispensable, valuable beings. Such population measures seek to control who can be a member of the population of indispensable beings and the population of dispensable beings globally. They are structural measures built into the world to produce and incorporate human beings in the Global North into the population of indispensable, valuable beings and human beings in Africa and other parts of the Global South into the population of disposable beings. Necroeconomic population control measures police procreation and ensure that procreation in Africa produces a population of disposable beings, and procreation in the Global North produces a population of indispensable and valuable beings.

7.4 The Racialized World Order and the Persistence of Population Control Policies

Now, I will show that the population control policies that promote the welfare of the population in the Global North and, at the same time, undermine the welfare of the population in the Global South are structural effects of a racialized world order. Coloniality of being is one of the constitutive principles that configure a racialized world, which is associated with othering non-Western populations as surplus disposable beings.[64] Coloniality of being is associated with ideologies that deny the humanity of Africans and that regard Africans as disposable non-beings. The notion of disposability connotes the destructive possibilities that involve the elimination of a being. The features that render a being disposable include the conditions of being deemed a surplus being and waste. A being that is constituted as surplus and as waste is disposable. Western racist thought constituted non-Western populations as surplus beings and as waste. Such a constitution of non-Western population groups as disposable surplus beings and waste began with the invention and deployment of a tool of coloniality, which Maldonado-Torres refers to as "racist imperial Manichaean misanthropic skepticism."[65] The "racist imperial Manichaean misanthropic skepticism" involves radical questioning and doubting of the humanity of Africans and other non-European population groups (Maldonado-Torres 2007, 245). Such "racist imperial Manichaean misanthropic skepticism" is related to what Boaventura de Sousa Santos refers to as abyssal thinking. Western abyssal thinking involves the use of invisible lines to dehumanize and expel other human beings from the category of the human.[66] The other human beings that are pushed into the "other

---

[64] Maldonado-Torres, N. 2007. On the coloniality of being. *Cultural Studies* 21: 240–270.

[65] Maldonado-Torres, On the coloniality of being, 245.

[66] See discussion of Boaventura de Sousa Santos understanding of abyssal thinking in Madlingozi, T. 2023. Social Movement Struggles for Decolonisation and (re)Constitution from Below:

side of the abyssal line" are invented as "invisible, unintelligible, or irreversibly discardable."[67] The questioning and rejection of the humanity of non-Western people involved a conception of humanity that differentiated the global human population into "inferior and superior, irrational and rational, primitive and civilized, traditional and modern" beings.[68] The victims of the Western abyssal thinking and the "racist imperial Manichean misanthropic skepticism" were "othered" as irrational, inferior, primitive, and premodern beings who are plagued by an ontological defect. The effect of the racist imperial Manichean misanthropic skepticism and abyssal thinking is the rejection of the humanity and value of non-Western people and their expulsion from the world of humans. The expulsion from the sphere of the human renders them superfluous beings and disposable waste.

The excoriation, denuding, and radical repudiation of the human status and value of non-Western people all pave the way for their construction/invention in multiple forms of disposability as a waste, threat, and superfluous populations. If disposable non-Western human beings are not wanted and are superfluous, then they can be disposed of as waste or eliminated. The two ways of treating a disposable surplus population of human beings as disposable waste, or as eliminable beings who can be destroyed, are the inevitable outcomes of "racist imperial Manichaean misanthropic skepticism" and abyssal thinking, which denied the humanity of Africans and other non-Western people.

Coloniality of being that othered non-Western people as disposable beings created a racialized world. Racialized necroeconomic population policy, which configures the economies of African countries to promote the welfare of the population in the Global North while simultaneously undermining the welfare of the population in the Global South and producing a population of disposable beings, emerges from a racialized world that is constituted by coloniality of being. The coloniality of being, which, as stated above, configures a racialized world order that produces Africans as disposable beings, was used to justify the necroeconomic configuration of the economies of Africa to promote the welfare of the population of the Global North and to undermine the welfare of the population of African countries. The idea that indigenous Africans are disposable waste is the constitutive racialized apparatus of population control that ensures that populations in the South suffer actual treatment as disposable beings.

The status of Africans as superfluous, disposable beings came to the fore during the European violent wars of colonial conquest that killed many in Africa and global

---

Abahlali baseMjondolo's Strivings against Pariahdom. In Research Handbook on Law, Movement and Social Change Eds. Steven A. Boutcher, Corey S. Shdaimah, and Michael W. Yarbrough, 1–26. Edward Elgar Publishing. 4. https://www.academia.edu/118636488/Social_Movement_Struggles_for_Decolonisation_and_re_Constitution_from_Below_Abahlali_baseMjondolos_Strivings_against_Pariahdom

[67] See discussion of Boaventura de Sousa Santos understanding of abyssal thinking in Madlingozi Social Movement Struggles 2023, 4.

[68] (Quijano cited in Ndlovu-Gatsheni, S. 2013. *Coloniality of Power in Postcolonial Africa: Myths of Decolonization.* Senegal: Codesria.16–17.

capitalist accumulation practices that impoverished and exploited Africans. Now, I will show how, during colonial wars of conquest, Africans were treated as disposable waste and were killed to eliminate resistance to colonial dispossession, looting of wealth, and plunder of resources.

The racialized colonialist and genocidal necroeconomic populationist measures were deployed to terminate the resistance to the Western capitalist exploitation of African natural resources and labour. According to Mueni wa Muiu and Guy Martin, the British colonial army sought a drastic reduction of the population of indigenous Africans by distributing blankets that were infected with smallpox to the Zulus in South Africa's Natal area.[69] Similarly, Indigenous populations of South Africa, such as the Khoikhoi and the San, were brutally exterminated by the British and Afrikaner colonialists.[70] Some of the extermination methods included deliberate inducement of epidemics. This led to the drastic decline of the population of the Khoikhoi, from 200,000 in 1652 to 20,006 in 1805.[71] The European invaders confined 115,700 indigenous Africans in refugee camps from June 1901 to May 1902.[72] 14,154 Africans died in the abovementioned camps.[73] In German South-West Africa (Namibia), General Lothar Von Trotha issued an extermination order for the Hereros on 2 October 1904.[74] Germany drove them into the desert and sealed the borders.[75] Such a ruthless action led to the death of 65,000 Heroros.[76] Additionally, there were massacres of vast African populations by the French in West Africa and Central Africa and by King Leopold in Congo.[77]

The Africans were killed to loot wealth in order to promote the welfare of populations in the Global North. One can say that the necroeconomic population policy of transferring wealth from Africa to the West by colonial looting not only undermined the welfare of Indigenous Africans to promote the welfare of the Western population but also involved the destruction of the lives of many human beings in Africa. Such a necroeconomic population policy could only emerge in a racialized world order that had been configured by coloniality of being, which others Africans as disposable beings. If Africans are othered as disposable beings in a racialized world, then their welfare and life can be undermined by a necroeconomic population policy to promote the well-being of the population of the Global North.

I have shown that necroeconomic population policy associated with undermining the lives of Africans to promote the welfare of the population of the Global North

[69] wa Muiu, M., and Martin, G. 2009. *A New Paradigm of the African State: Fundi wa Afrika* New York: Palgrave Macmillan. 65.

[70] wa Muiu, M., and Martin, G. *A New Paradigm of the African State.*

[71] wa Muiu, M., and Martin, G. *A New Paradigm of the African State.* 67.

[72] wa Muiu, M., and Martin, G. *A New Paradigm of the African State.* 67.

[73] wa Muiu, M., and Martin, G. *A New Paradigm of the African State.* 67.

[74] wa Muiu, M., and Martin, G. *A New Paradigm of the African State.* 69.

[75] wa Muiu, M., and Martin, G. *A New Paradigm of the African State.* 69.

[76] wa Muiu, M., and Martin, G. *A New Paradigm of the African State.* 69.

[77] wa Muiu, M., and Martin, G. *A New Paradigm of the African State.*

occurred in a racialized world constituted by global coloniality, which othered Indigenous Africans as disposable beings. If necroeconomic population policy originates in a racialized world order that is constituted by coloniality of being, then the condition for the termination of such population policy entails the liquidation of a racialized world order. The racialized world was not abolished by the decolonization movement that terminated the direct colonial administration of countries in Africa. Anti-colonial national liberation movements did not dismantle a racialized world order. Grosfoguel holds that decolonization achieved juridical political independence.[78] However, the racialized capitalist world order remained.[79] There was truncated decolonization, which sought the integration of the Global South people, who were constituted as disposable beings, into a racialized world order. If the necroeconomic population policy is a structural effect of a racialized world that is configured by coloniality of being, then the termination of such a population policy is dependent on the liquidation of a racialized world order. The failure of the decolonization movement to terminate a racialized world order perpetuated racialized population policies such as necroeconomic population policy.

We can situate population control measures that persist, and that survived the truncated and pretentious decolonization within a racialized world order. It is the racialized world order that reproduces population control policies. The Cairo Conference focused on coercive population control policies that sought to reduce the fertility of poor, marginalized women in the Global South. However, the meeting did not consider the idea that a racialized world order reproduces population control policies that structurally control the production of populations of disposable beings in the Global South. The idea of disposable beings is associated with an undermining of the welfare of human beings in Africa to promote the welfare of the population in the Global North. Additionally, population control policies such as biopopulationism, demopopulationism, geopopulationism, and necropopulationism continued after the Cairo Conference, which had sought to terminate just such coercive population control policies.[80] Additionally, the population policies that sought to reduce the fertility rates of women in the Global South were reconfigured and reframed as part of the discourse that promotes reproductive rights and health.[81] Such population policies, which seek to reduce birth rates, are harmful demographization tools that reduce the problem of poverty in the Global South to a demographic problem of high birth rates and overpopulation, which causes resource strain.[82] This conceals the racialized necroeconomic ordering of the African economies that promotes the welfare of the population of the Global North while at the same time causing poverty and undermining the welfare of the African population.

---

[78] Grosfoguel, R. 2011. Decolonizing Post-Colonial Studies and Paradigms of Political Economy: Transmodernity, Decolonial Thinking, and Global Coloniality. *Transmodernity: Journal of Peripheral Cultural Production of the Luso-Hispanic World* 1:1–39.

[79] Grosfoguel, R. Decolonizing Post-Colonial Studies.

[80] Bhatia et al. A feminist exploration of 'populationism'.

[81] Bhatia et al. A feminist exploration of 'populationism'.

[82] Schultz, S. Demographic futurity: How statistical assumption politics shape immigration policy.

The Cairo Conference could not terminate population control as it did not seek the termination of a racialized world that reproduces population control policies. Also, the Cairo Conference did not deal with the broad population control policies, such as necroeconomic population policies that control economies in Africa and transfer wealth from Africa to promote the well-being of the population in the Global North, while at the same time undermining the welfare of many Africans and producing a population of disposable beings in Africa.

Racialized population control policies are a structural effect of a racialized world order. If global coloniality is the architectural principle for configuring the world, then we can have racialized population policies that target fertility control of poor women from the Global South. The necroeconomic population control policy is an inevitable global structural tool for promoting the welfare of the Global North, which is tied simultaneously to undermining the welfare of the population in the Global South. The trampling of the welfare of the population in the Global South is a necessary, permanent, and ineradicable feature of the promotion of the welfare of the population of the Global North in a racialized world. Population control measures such as necroeconomic population policy can only be terminated by obliterating a racialized world order.

7.5 Malthusian Thinking, Thinking that Renders Humanity Abstract and Relative

One can contend that Malthusian thinking and the thinking that renders humanity "abstract and relative" is the source of population control.[83] Such thinking involves a quantitative relationship between human beings and the economy, as well as between human beings and the environment.[84] If there is a disequilibrium, then the solution can be fertility control. This leads to targeting racialized populations in the global south with fertility control. There is an underlying logic that the main problem is Malthusian thinking, an approach that renders humanity abstract and relative. If we eliminate Malthusian thinking, and thinking that renders humanity abstract and relative, then we can also eliminate population control and the targeting of racialized populations in the Global South with population control.

The thinking that renders humanity relative and abstract, and Malthusian thinking, results from a world plagued by poverty, environmental degradation, and climate change. If there were no problems of poverty, environmental degradation, and climate change, then we could not have Malthusian thinking and such problematic thinking that renders human beings relative and abstract. The solution is to terminate a world configured by capital and race to produce poverty largely in the Global

[83] See Bhatia et al. Bhatia et al. A feminist exploration of 'populationism'. 4. on the kind of thinking that make humanity "abstract" and "relative."

[84] Bhatia et al. A feminist exploration of 'populationism'.

South, the climate crisis, and environmental degradation. Marxist feminist political and ecological scholars, such as Diana Ojeda, Jade Sasser, and Elizabeth Lunstrum, hold that environmental degradation and climate change result from capitalist overexploitation and overconsumption by a few rich elites, mainly from the Global North and the emerging economies.[85] If the world is constituted by capital and race to produce poverty, climate crisis, and environmental degradation, then the thinking that makes human beings relative and Malthusian thinking are symptoms of a faulty and dysfunctional world constituted by race and capital. In addition, we cannot regard Malthusian thinking as the cause of racialized population control that targets people in the Global South with fertility control. Racialized population control is not an effect of Malthusian thinking. It is an effect of the racialized world order. Malthusian thinking took a racialized form of fertility control in the Global South as a false panacea to environmental degradation and climate change in a racialized world order. It did not evolve and take on a racialized form of its own accord. It took a racialized form when the modern capitalist world was configured by coloniality of being. The racialized form of Malthusian thinking is a result of a racialized world; the classed form is a result of the configuration of the world by capital. Also, racialized uneven global capitalist development was configured by capital and race to produce a huge class of racialized poor disposable beings in Africa. It is such a racialized poor population that is targeted by Malthusian thinking. Malthusian thinking in a racialized capitalist world order is a structural effect of capital and race. Malthusian thinking is not the source of racialized and classed population control. Rather, racialized and classed population control and Malthusian thinking are effects of a racialized and capitalist world. The termination of racialized and classed population control requires the termination of a racialized and capitalist world order. The existence of Malthusian thinking and other forms of thinking that spawn population control is highly unlikely in a non-capitalist and deracialized world order that promotes the welfare of all human beings.

7.6 Welfare and Technology

I have shown that there is a necroeconomic population policy that promotes the welfare of the population in the Global North while simultaneously undermining the welfare of the population in the Global South. One can argue that the welfare of the population of the Global North is promoted by the technological innovation associated with the first, second, third, and fourth industrial revolutions. The increase in the population density of Europe was associated with the better

[85] Ojeda, D. Sasser, J. D. and Lunstrum, E. 2019. Malthus's Specter and the Anthropocene. *Gender, Place & Culture* 1–17.

organization of labour, specialization, and technological innovation, which maximized the use of land and resources to promote the welfare of the population.[86]

However, such a view, which maintains that technological innovation is at the core of the promotion of the welfare of the population of the Global North, is problematic, as all the technological innovations associated with all the industrial revolutions involved necroeconomic practices that undermined the welfare of the population of Africa, while at the same time promoting the welfare of the population in the Global North. The technological innovation associated with the mechanization of production during the first industrial revolution was accompanied by the transatlantic slave trade.[87] The technological innovation associated with the invention of electricity during the Second Industrial Revolution, which led to mass production, was accompanied by colonialism, colonial domination, and the looting of the wealth of Africans.[88] Moreover, the Industrial Revolution associated with information and communication technology innovations was accompanied by neoliberal capitalist domination, which involved the loss of economic sovereignty for many African countries. Economic policies such as economic liberalization and privatization, which favoured economic exploitation by foreign multinational corporations, were dictated by the International Monetary Fund (IMF) and the World Bank. Also, the artificial intelligence (AI) technological innovation associated with the Fourth Industrial Revolution is accompanied by green imperialism in Africa and other parts of the Global South, which involves the exploitation of energy transition minerals by foreign multinational corporations. It is questionable to suggest that technological innovations in the Global North are the primary factors that account for the promotion of the welfare of the population of the Global North.

7.7 Conclusion

I have shown that necroeconomic population policy functions to promote the welfare of the population of the Global North while at the same time undermining the welfare of the population of Africa and the other parts of the Global South. I have also shown that necroeconomic population policy and other population policies are reproduced and sustained by a racialized world order that is configured by coloniality of being, which "others" Africans and other non-Western people as disposable beings. I have shown that the termination of population control policies is contingent on the annihilation of the racialized capitalist world. Future research should

[86] Kolk, M. 2022. Demographic Theory and Population Ethics. In *The Oxford handbook of Population Ethics*. Eds. Gustaf Arrhenius, Gustaf Krister Bykvist, Tim Campbell and Elizabeth Finneron-Burns,468–489. Oxford: Oxford University Press.

[87] Benyera, E. 2022. On the Elusiveness of Epistemic Freedom in Africa and the Need to Unmask the Colonial Contract. In *Africa and the Fourth Industrial Revolution: Curse or Cure?* Ed. Everisto Benyera, 1–15. Cham: Springer.

[88] Benyera, On the Elusiveness of Epistemic Freedom.

shed further light on the revolutionary agents for abolishing a racialized and capitalist world and for configuring a new world that promotes the welfare of all human beings and that eschews racialized population control.

References

Amin, S. 1990. *Delinking: Towards a Polycentric World, translated by Michael Wolfers*. London: Zed Books.

Benyera, E. 2022. On the elusiveness of epistemic freedom in Africa and the need to unmask the colonial contract. In *Africa and the fourth industrial revolution: Curse or cure?* ed. Everisto Benyera, 1–15. Cham: Springer.

Bhatia, R., S. J. Sasser, D. Ojeda, A. Hendrixson, S. Nadimpally, and E. E. Foley. 2019. A feminist exploration of 'populationism': Engaging contemporary forms of population control. *Gender, Place & Culture* 27:333–350.

Bond, P. 2019. Luxemburg's critique of capital accumulation, reapplied in Africa. *Journal Für Entwicklungspolitik* XXXV:92–118.

Grosfoguel, R. 2011. Decolonizing post-colonial studies and paradigms of political economy: Transmodernity, decolonial thinking, and global coloniality. *Transmodernity: Journal of Peripheral Cultural Production of the Luso-Hispanic World* 1:1–39.

International Labour Report. 2024. *Global employment trends for youth 2024 Sub-Saharan Africa*. https://www.ilo.org/sites/default/files/2024-08/Sub-Saharan%20Africa%20GET%20Youth%202024_0.pdf

Kolk, M. 2022. Demographic theory and population ethics. In *The Oxford handbook of Population Ethics*, ed. Gustaf Arrhenius, Gustaf Krister Bykvist, Tim Campbell, and Elizabeth Finneron-Burns, 468–489. Oxford: Oxford University Press.

Madlingozi, T. 2023. Social movement struggles for decolonisation and (re)constitution from below: Abahlali base Mjondolo's strivings against Pariahdom. In *Research handbook on law, Movement and social change*, ed. Steven A. Boutcher, Corey S. Shdaimah, and Michael W. Yarbrough, 1–26. Edward Elgar Publishing. https://www.academia.edu/118636488/Social_Movement_Struggles_for_Decolonisation_and_re_Constitution_from_Below_Abahlali_base-Mjondolos_Strivings_against_Pariahdom.

Maldonado-Torres, N. 2007. On the coloniality of being. *Cultural Studies* 21:240–270.

Mbembe, A. 2019. *Necropolitics. Translated by Steven Corcoran*. Durham and London: Duke University Press.

Ndlovu-Gatsheni, S. 2013. *Coloniality of power in postcolonial Africa: Myths of decolonization*. Senegal: Codesria.

Ojeda, D., J. D. Sasser, and E. Lunstrum. 2019. Malthus's specter and the anthropocene. *Gender, Place & Culture*1–17.

Radley, B. 2023. Green imperialism, sovereignty, and the quest for national development in the Congo. *Review of African Political Economy* 50 (177–178): 322–339.

Richey, L. A. 2010. Reproductive health, family planning, and HIV/AIDS: Dangers of (Dis) Integration in Tanzania and Uganda. In *Markets and malthus population, gender, and health in neo-liberal times*, ed. Mohan Rao and Sarah Sexton, 265–298. New Delhi: Sage Publications.

Schultz, S. n.d. *Demographic futurity: How statistical assumption politics shape immigration policy rationales in Germany*. https://www.academia.edu/36758617/Demographic_futurity_How_statistical_assumption_politics_shape_immigration_policy_rationales_in_Germany_2018_

Shaw, A., and K. Wilson. 2020. The bill and melinda gates foundation and the necro-populationism of 'climate-smart' agriculture. *Gender, Place & Culture* 27:370–393.

Wa Muiu, M., and G. Martin. 2009. *A new paradigm of the African State: Fundi wa Afrika*. New York: Palgrave Macmillan.
Wilson, K. 2017. Re-centring 'race' in development: Population policies and global capital accumulation in the era of the SDGs. *Globalizations* 14:432–449.
Yates, M. 2011. The human-as-waste, the labor theory of value and disposability in contemporary capitalism. *Antipode* 43:1679–1695.
UNCTAD. 2022. *World investment report*. New York: United Nations Publications.
UNCTAD. 2019. *Key statistics and trends in regional trade in Africa*. Geneva: United Nations.

Chapter 8
Rethinking Population Ethics and Environmental Sustainability through African Communitarian Thought

Stephen Nkansah Morgan

Abstract This chapter explores the ethical imperative of controlling the human population for environmental sustainability through the lens of African communitarian ethics. It begins by situating the discourse within broader Western debates on population ethics, including utilitarian, deontological, person-affecting, and rights-based approaches, while highlighting their contributions and limitations, especially in addressing intergenerational justice, environmental degradation, and the moral status of potential persons. The chapter then turns to arguments for and against population control, noting both the environmental necessity of reducing human pressure on planetary resources and the moral risks of coercive or inequitable policies. It argues that African communitarianism, with its emphasis on interconnectedness, collective responsibility, cultural embeddedness, and care for future generations, offers a holistic and ethically grounded alternative.

Keywords Ecological collective well-being · Environmental sustainabilty · Population control · African communitarian ethics

8.1 Introduction

In recent decades, the human population has grown at an unprecedented rate, triggering profound impacts on the environment and raising urgent ethical questions about issues of sustainability, justice, and responsibility. As of 2025, the global population stands at approximately 8.1 billion, with projections by the United Nations suggesting a rise to nearly 9.7 billion by 2050 and over 10.4 billion by the

S. N. Morgan (✉)
Department of Philosophy and Classics, University of Ghana,
Accra, Greater Accra Region, Ghana
e-mail: smorgan@ug.edu.gh

B. Okyere-Manu, P. Mwipikeni (eds.), *An African Approach to Population Ethics*, Advancing Global Bioethics 23,
https://doi.org/10.1007/978-3-032-24464-2_8

end of the century if current trends persist.[1] This rapid growth is especially pronounced in parts of Africa, where equivalent declines have not matched improvements in healthcare and declining mortality in fertility. While population growth can reflect positive developments such as better child survival and improved public health, its environmental consequences are undeniable, from increasing greenhouse gas emissions and freshwater depletion to deforestation and biodiversity loss.

One visible example of the strain posed by rapid population growth is to be found in Lagos, Nigeria, one of the fastest-growing cities in the world. With an estimated population exceeding 22 million and growing at a rate of over 3% annually, Lagos faces severe challenges in providing adequate housing, clean water, sanitation, and sustainable infrastructure.[2] The city's wetlands are disappearing under the pressure of urban expansion, and its waste management systems are overwhelmed, contributing to widespread pollution and ecological degradation. These issues are not unique to Lagos, of course; they are emblematic of broader tensions between human population pressures and environmental sustainability, especially in urban centres across the Global South.

In response to these challenges, global discourse has increasingly turned toward the ethics of population control. Yet, this discourse remains ethically contentious, often dominated by Western philosophical frameworks such as utilitarianism, deontology, and rights-based theories. These frameworks, while insightful, tend to prioritize individual autonomy or abstract calculations of harm, sometimes overlooking the communal dimensions of ethical life. Moreover, historical abuses associated with population control, such as coercive sterilizations, racist policies, and paternalistic interventions, have left deep scars and fostered understandable skepticism toward the idea of regulating human reproduction.

This chapter offers a different ethical lens, an African communitarian ethical perspective, for engaging the complex relationship between population and environmental sustainability. Rooted in African worldviews that prioritize relationality, community, and collective well-being, African communitarian thought provides a moral framework that is deeply attentive to both human dignity and ecological balance. As John Mbiti famously stated, "I am because we are, and since we are, therefore I am."[3] This conception of personhood situates individuals within a network of social and ecological relationships, implying that ethical decisions, including those about reproduction and population, must account for their impact on the wider community and environment.

The chapter draws on key philosophical insights, including Kwame Gyekye's idea of the common good, which emphasizes the shared interests and well-being of

[1] From *World Population Prospects 2022. Summary of Results*, by United Nations, 2022, United Nations Population Division, Available online at https://population.un.org/wpp/.

[2] From *Lagos Urban Transport and Environmental Assessment Report*, by World Bank, 2021, World Bank Publications, Available online at https://documents1.worldbank.org/curated/en/309711468031496273/pdf/WPS5730.pdf

[3] From *African Religions and Philosophy*, by J. S. Mbiti, 1990, Heinemann.

society as a whole.[4] In the context of population ethics, this means that reproductive choices are not solely private matters but also ones of collective concern, especially when they have far-reaching environmental implications. Similarly, Godfrey Tangwa's principle of "*live-and-let-live*" extends the moral community to include nonhuman animals and the broader ecology.[5] Tangwa's ethic calls for a way of living that makes room for all life forms to flourish. Achieving this, he argues, may require moderating human population growth—not through coercion, but through ethically grounded policies and community-based approaches that promote voluntary and informed family planning.

The chapter is structured into seven main sections. Following this introduction, the next section provides a brief historical background and ethical critique of Western population control discourse. This is followed by an exploration of population ethics and environmental sustainability in Western literature. Here, the chapter considers key conceptual clarifications of population, population control, and environmental sustainability. It also explores the ethical interrelationship between population and environmental sustainability, setting the foundation for later arguments. The next section outlines and critically examines major arguments for and against population control, highlighting both the ecological rationale and the ethical tensions involved. The chapter then presents and develops an African communitarian ethical perspective on the issue, drawing on thinkers such as Mbiti, Gyekye, Tangwa, and others. Finally, the chapter anticipates possible critiques and offers reasoned responses, particularly addressing concerns about coercion, rights, and cultural sensitivity. The chapter concludes with a reflective synthesis that underscores the value of African communitarian ethics in shaping equitable and sustainable population ethics and environmental sustainability.

8.2 Historical Background and Ethical Critique of Western Population Control Discourse

The global conversation on population control has a long and complex history, much of it shaped by Western ideas about development, progress, and environmental limits. Since the late eighteenth century, thinkers like Thomas Malthus[6] have argued that unchecked population growth would outpace food production, leading to poverty, famine, and social unrest. His theory, known as Malthusianism, has had a lasting influence on both public policy and popular thought, reinforcing the idea that population growth is a direct threat to human well-being and ecological stability.

---

[4] From *Tradition and Modernity: Philosophical Reflections on the African Experience*, by K. Gyekye, 1997, Oxford University Press.

[5] From Some African Reflections on Biomedical and Environmental Ethics, by G. B. Tangwa, In *A Companion to African Philosophy*, 387–395, by Kwasi Wiredu (ed.), 2004, Blackwell.

[6] T. R. Malthus, *An Essay on the Principle of Population*, 1978. J. Johnson, London.

In the twentieth century, these ideas were revived and reshaped through neo-Malthusian perspectives, particularly during the post-World War II period when concerns about global overpopulation gained international attention. Influential works such as *The Population Bomb* by Paul Ehrlich[7] warned of imminent ecological collapse if population growth was not curbed. These arguments led to large-scale interventions by governments and international organizations, particularly in the Global South, including in Asia, Latin America, and parts of Africa. These interventions often included aggressive family planning campaigns, forced sterilizations, and financial incentives or penalties tied to reproductive behavior.

While some of these programs did lead to reductions in fertility rates, they also raised serious ethical questions. Many policies ignored cultural contexts, devalued women's agency, and prioritized population numbers over human dignity. Scholars have criticized these approaches for treating human beings as problems to be solved, rather than as moral agents embedded in social and cultural worlds.[8,9] Furthermore, critics have pointed out that population control measures often targeted poor, racialized, and indigenous communities, reinforcing global inequalities and echoing colonial patterns of control.[10]

Another major criticism of Western population discourse is that it often reduces environmental problems to issues of human numbers, rather than focusing on patterns of consumption, production, and inequality. High-consuming countries with relatively low population growth contribute disproportionately to environmental degradation, yet population control efforts have mostly focused on low-income countries. This creates a double standard and shifts attention away from more pressing causes of environmental harm, such as fossil fuel dependence, industrial agriculture, and unsustainable economic models.[11]

In recent years, some development and environmental organizations have tried to move toward a more rights-based approach to population and reproductive health. However, the legacy of coercive policies and technocratic thinking continues to shape how the population is discussed and managed globally. This chapter argues that a deeper shift is needed, one that not only respects human rights but also recognizes alternative ethical traditions that can guide population and environmental decision-making in more holistic and culturally grounded ways.

By turning to traditional African communitarian ethics, we can begin to rethink what it means to live sustainably, in terms not just of numbers or resources, but also

---

[7] *The Population Bomb* by P. R. Ehrlich, 1968, Ballantine Books.

[8] See for example, *Reproductive Rights and Wrongs: The Global Politics of Population Control* by B. Hartmann, 1995, South End Press.

[9] See for example *Dangerous Intersections: Feminist Perspectives on Population, Environment, and Development* by J. Silliman & Y. King (Eds.), 1999, South End Press.

[10] From *Fatal Misconception: The Struggle to Control World Population*, by M. Connelly, 2008. Harvard University Press.

[11] See *The Demography of Adaptation to Climate Change*, by G. Martine, & D. Schensul, 2013, UNFPA, IIED.

of relationships, responsibilities, and the moral values that guide human life within the natural world.

8.3 Population Ethics and Environmental Sustainability in Western Discourse

Population ethics explores the ethical dimensions of decisions that affect population size, composition, and the well-being of current and future individuals. It asks fundamental questions: Is it better to bring more people into existence, or to improve the quality of life for those who already exist? How should we balance the needs of present populations with those of future generations? These questions are crucial in addressing contemporary global challenges such as climate change, biodiversity loss, and resource depletion. In Western discourse, population ethics intersects with environmental sustainability debates in ways that have significantly shaped global policy and academic thinking.

Western approaches to population ethics are grounded in three main ethical frameworks: utilitarianism, deontology, and person-affecting views. Each of these offers distinct insights and challenges.

8.3.1 Utilitarian Approaches

Utilitarianism, closely associated with philosophers such as Jeremy Bentham and John Stuart Mill, holds that the morality of actions is determined by their consequences, and specifically by the amount of happiness or utility they produce. In population ethics, utilitarianism manifests in two dominant forms: total utilitarianism and average utilitarianism.

Total utilitarianism argues that we should seek to maximize the total sum of well-being in a population. Derek Parfit,[12] a key figure in population ethics, noted that this view leads to what he termed the "Repugnant Conclusion," the idea that a massive population living lives barely worth living could be preferable to a smaller population enjoying very high well-being, as long as the total level of happiness is greater.

Average utilitarianism, by contrast, seeks to maximize the average well-being per person. This view avoids the "Repugnant Conclusion," but has its problems. For example, it could imply that reducing the number of less happy people (even through morally questionable means) is ethically desirable, as this would increase the average happiness.[13]

[12] D. Parfit, *Reasons and Persons*, 1984, Oxford University Press.

[13] See *Counting the Cost of Global Warming*, by J. Broome, 1992, Cambridge, White Horse Press.

In environmental discourse, utilitarian models have often underpinned arguments for controlling population growth as a means of optimizing resource use and reducing ecological strain. However, these models tend to reduce individuals to units of utility and populations to calculable risks, overlooking cultural, social, and moral dimensions of population life.

8.3.2 Person-Affecting and Deontological Views

Person-affecting views, articulated by philosophers like Jan Narveson[14] and Jeff McMahan,[15] argue that moral actions are those that affect specific individuals. According to this view, bringing someone into existence cannot be good or bad for that person, since they would not otherwise exist. These views focus moral concern on those who already exist, raising difficult questions about intergenerational ethics and environmental responsibility.

Some person-affecting theorists endorse the *asymmetry view*, which holds that it is wrong to bring into existence a person who will suffer, but not necessarily right to bring into existence someone who will be happy. While these perspectives respect individual dignity, they can struggle to account for our obligations to future generations, especially in the face of long-term environmental degradation.

Deontological ethics, associated with Immanuel Kant, emphasizes moral duties and the intrinsic worth of individuals. From this perspective, actions must be guided by moral principles rather than their outcomes. Applied to population ethics, deontology stresses reproductive rights, autonomy, and the moral imperative not to treat people as mere means to an end.[16] This view challenges utilitarian population control policies that may violate individual rights for collective benefit.

As environmental crises worsen, rights-based approaches, often inspired by deontological ethics, have gained prominence. These perspectives argue that we have moral duties not only to current but also to future generations. John Rawls,[17] for instance, emphasized the principle of justice between generations, insisting that policies should ensure that future people have access to the same opportunities and resources as current ones.

Western population ethics, while rich in theoretical depth, often struggles with real-world complexities. Utilitarian models risk endorsing coercive or reductionist policies, person-affecting theories face limitations in addressing intergenerational justice, and deontological views may lack the flexibility needed to address global-scale problems. Moreover, the dominant discourse tends to frame environmental

[14] J. Narveson, Moral Problems of Population, 1973, *The Monist*, 57, 62–86.

[15] J. McMahan, Problems of Population Theory, 1981, *Ethics*, 92 (1), 96–127.

[16] See *Groundwork for the Metaphysics of Morals*, by I. Kant, 1785, Cambridge University Press.

[17] Rawls, J. *A Theory of Justice*, 1971, Cambridge, MA. Harvard University Press.

challenges in terms of numbers and abstract calculations, marginalizing the moral and spiritual relationships that people have with the land, community, and future life.

These limitations reveal the need for alternative perspectives, ones that embed ethical reflection within relational, communal, and ecological contexts. This chapter proposes that traditional African communitarian ethics offer such a framework. Rooted in values of interdependence, respect for life, and harmony with nature, African communitarian thought challenges the individualistic and often-technocratic orientation of Western population discourse. It offers a more holistic ethical basis for addressing population and environmental sustainability in a way that centres human dignity and ecological integrity.

8.4 Arguments for and Against Population Control

Population, broadly defined, refers to the total number of individuals of a particular species residing in a specific geographic area at a given time. In human contexts, population studies intersect with disciplines such as ecology, sociology, and demography to explore how population size, growth, and distribution influence and are influenced by environmental systems. One key concern in present-day environmental ethics today is the relationship between human population dynamics and ecological sustainability.

The interaction between population size, affluence, and technology was famously captured in the equation $I = P \times A \times T$ by Ehrlich and Holdren,[18] where environmental Impact (I) is a function of Population (P), Affluence (A) (consumption per capita), and Technology (T) (impact per unit of consumption). This framework has been central to debates on whether limiting population growth is necessary to reduce environmental degradation. However, such discussions are ethically and politically charged and have generated both compelling support and serious criticism.

8.4.1 Arguments for Population Control

1. Resource Allocation and Environmental Impact

Proponents argue that unchecked population growth leads to increased demand for finite resources, such as water, food, energy, and land, resulting in resource depletion, habitat destruction, and pollution. As populations expand, the strain on ecosystems intensifies, often beyond their capacity to regenerate. From this standpoint, population control is seen as a necessary measure to reduce human pressure on the environment and promote long-term ecological balance.

[18] P. R. Ehrlich, & J. P. Holdren, Impact of Population Growth, 1971, *Science*, 171 (3967), 1212–1217.

2. Climate Change Mitigation

Population growth is also linked to increased greenhouse gas emissions, especially in contexts where both industrialization and urbanization are accelerating. A growing body of research indicates that population stabilization could significantly reduce future carbon emissions.[19] Advocates argue that limiting population growth, particularly in high-emitting regions, could be a cost-effective strategy for achieving climate goals and slowing global warming.

3. Support for Sustainable Development Goals (SDGs)

Global development frameworks, such as the United Nations Sustainable Development Goals, indirectly support population policies that promote environmental sustainability. For example, Goal 11, which aims to make cities inclusive, safe, resilient, and sustainable, is difficult to achieve in contexts of rapid, unplanned population growth. Advocates suggest that access to voluntary family planning, education, and women's empowerment can contribute to both population stabilization and broader development outcomes.

4. Ecological Stewardship and Land Ethics

Some environmental philosophers, such as Aldo Leopold,[20] have called for a "land ethic" that recognizes the interconnectedness of all living things and the need to limit human encroachment on natural systems. From this perspective, population control is not just a pragmatic solution, but a moral obligation, a way to live in harmony with nature and to protect the biosphere from collapse.

8.4.2 Arguments Against Population Control

1. Ethical Concerns and Human Rights

One major critique of population control policies is their potential to violate individual rights, particularly reproductive autonomy. Historical cases of coercive sterilizations and punitive birth policies, especially in marginalized communities, have raised red flags about the ethical implications of state-led or externally imposed population measures.[21,22] Critics argue that reproductive choices must be guided by individual consent and supported through access to education and healthcare, not through coercion or restriction.

2. Misplaced Focus: Consumption and Global Inequality

---

[19] See for example, Population Stabilisation and Climate Change Mitigation: A Global Perspective, by S. D'Alessandro, et al., 2017, *Environmental Science & Policy*, 77, 1–10.

[20] A. Leopold, *A Sand County Almanac*, 1949, Oxford University Press.

[21] See Experiences of forced sterilization and coercion to sterilize among women living with HIV (WLHIV) in Namibia: an analysis of the psychological and socio-cultural effects, by K. Bakare, & S. Gentz, 2020, *Sexual and reproductive health matters*, 28 (1), 1758439. https://doi.org/10.1080/26410397.2020.1758439

[22] Also see *Fatal Misconception: The Struggle to Control World Population*, by M. Connelly, 2008, Harvard University Press.

Many environmental ethicists and justice advocates argue that overpopulation is a misleading or partial diagnosis of environmental problems. They point out that the wealthiest 10% of the global population contribute disproportionately to environmental degradation through high consumption patterns, while poorer communities, often those with higher birth rates, have a relatively small ecological footprint.[23] From this view, efforts to reduce population without addressing overconsumption and systemic inequality risk scapegoating the poor while allowing the affluent to avoid accountability.

3. Technological Optimism and Innovation

Opponents of population control also emphasize the role of technology in addressing environmental challenges. They argue that human ingenuity has historically increased the Earth's carrying capacity through innovations in agriculture, energy, and waste management. From this perspective, they argue that, rather than limiting human numbers, we should invest in sustainable technologies and systems that decouple economic growth from environmental harm.[24] This approach foregrounds adaptive capacity and problem-solving rather than restriction.

4. Cultural and Social Dimensions

Population control debates are often framed in universal or technocratic terms, neglecting the cultural, historical, and ethical diversity of human societies. Critics note that dominant population discourses tend to privilege Western, individualistic perspectives, while marginalizing Indigenous and non-Western worldviews that emphasize community, relationality, and coexistence. This universalizing tendency risks imposing solutions that may be inappropriate or harmful in diverse cultural contexts.

8.5 An African Communitarian Ethical Perspective on Population and Environmental Sustainability

African communitarian ethics offers a culturally rooted framework for thinking about the ethical challenges related to population growth and environmental sustainability. It emphasizes collective well-being, shared responsibility, and the interconnectedness of all life—values that stand in contrast to the individualistic and often-anthropocentric tendencies in dominant Western ethical paradigms. This worldview situates the human person within a complex web of social, ecological, and spiritual relationships, where decisions, especially those related to

[23] See for example in Growth in Emission Transfers via International Trade from 1990 to 2008, by G. P. Peters, et al., 2011, *Proceedings of the National Academy of Sciences*, 108 (21), 8903–8908.

[24] See for example in Advanced Technology Paths to Global Climate Stability: Energy for a Greenhouse Planet, by M. I. Hoffert, et al., 2002, *Science*, 298 (5595), 981–987.

reproduction, resource use, and sustainability, must be evaluated in terms of their effects on the community and the broader environment.

A foundational idea in African communitarian thought is that personhood is not merely an individual attribute, but rather something achieved through relationships within the community. Mbiti[25] famously articulated this view in the phrase, "I am because we are; and since we are, therefore I am." This relational conception of the self implies that ethical decisions about population, such as family planning or the use of communal resources, must be grounded in considerations of the collective good rather than individual preference alone. In this framework, population policies should be assessed based on their contribution to communal harmony, sustainability, and the continuation of life across generations.

African communitarianism also places significant emphasis on collective responsibility. Individuals are morally accountable not only for their personal choices but also for how those choices affect others in the community. Wiredu[26] notes that many African societies prioritize actions that promote social cohesion and communal welfare. Reproductive choices, for instance, are not seen solely as private matters but as decisions with wider implications for the stability, resource distribution, and sustainability of the group. This view supports non-coercive, community-driven approaches to population management that encourage informed choices aligned with the needs of the entire community.

A key strength of African communitarian ethics lies in its attentiveness to cultural context and ethical pluralism. Different communities have varying beliefs about family size, reproductive norms, and environmental stewardship. As Eze[27] argues, ethical decisions must be responsive to these local realities. Imposing universal population policies without attention to cultural traditions can lead to alienation, resistance, or injustice. Instead, ethical frameworks should engage with the values, customs, and lived experiences of communities, fostering dialogue about how best to align traditional values with modern challenges.

For example, many African cultures uphold the extended family system, shared child-rearing, and strong intergenerational ties. These values offer both opportunities and constraints for developing ethical responses to population growth. When harnessed thoughtfully, they can become powerful resources for promoting sustainability through collective action and support for ethical reproductive choices.

The concept of *Ubuntu*, often translated as "a person is a person through other persons," extends ethical responsibility not only to current community members but also to future generations. Ubuntu ethics invites people to think long-term and to act with regard for the survival and flourishing of others, including those who have not yet been born.[28] This orientation supports sustainable development and population

[25] From *African Religions and Philosophy*, by J. S. Mbiti, 1990, Heinemann.

[26] From *Cultural Universals and Particulars: An African Perspective*, by K. Wiredu, 1996, Indiana University Press.

[27] From What is African communitarianism? Against consensus as a regulative ideal, by M. O. Eze, 2008, *South African Journal of Philosophy*, 27 (4), 386–399.

[28] See *No Future Without Forgiveness*, by D. Tutu, 1999, New York: Doubleday.

policies that protect the ability of future generations to access land, water, food, and a habitable environment.

Several African philosophers and ethicists note that the continent's traditions already contain embedded norms of ecological care and restraint.[29] From this standpoint, ethical population policies are not about restricting human freedom, but about exercising moral foresight to ensure that present actions do not undermine the conditions for continued communal life.

Justice and equity are also vital to the African communitarian perspective. Population policies must be just, not only in intention but also in implementation. They must avoid exacerbating existing inequalities or disproportionately burdening vulnerable groups. Gyekye[30] articulates a robust conception of the *common good* in Akan ethics, which he defines as the shared conditions necessary for individuals to flourish within the community. These include access to food, healthcare, education, and environmental resources. The goal of ethical governance, including in population matters, is to safeguard this common good.

This approach avoids framing population control as a technical or demographic problem. Instead, it situates it within a moral economy where what matters is ensuring that all people have the opportunity to live dignified lives in harmony with one another and with nature. Population management strategies must therefore aim at promoting equity in access to reproductive health services, education, and livelihood opportunities, especially for women and marginalized groups.

Cameroonian philosopher Godfrey Tangwa[31] offers a further deepening of African environmental ethics with his concept of *live-and-let-live.* This principle extends moral concern beyond humans to include nonhuman animals and ecosystems. According to Tangwa, every form of life has a right to exist and to flourish. Human beings, as the most powerful actors in the ecological web, bear a special responsibility to ensure that their activities do not crowd out or destroy other forms of life.

From this standpoint, unchecked human population growth poses a serious ethical concern. Not because it leads to too many people per se, but because it often translates into overuse of resources, destruction of habitats, and marginalization of other species. Tangwa's ethics call for self-restraint and respect for all life forms. Population control, understood as a moral duty to maintain ecological balance, is part of what it means to "live and let live." However, this must not involve the elimination of existing human beings or policies that violate human dignity. Rather, it

[29] See for example in Moral obligations towards future generations in African thought, by K. G. Behrens, 2012, *Journal of Global Ethics*, 8(2–3), 179–191. https://doi.org/10.1080/17449626.2012.705786

[30] From Person and community in African thought, by K. Gyekye, In *Person and community: Ghanaian philosophical studies I*, 101–122, by K. Wiredu & K. Gyekye (Eds.), 1992. Council for Research in Values and Philosophy.

[31] From Some African Reflections on Biomedical and Environmental Ethics, by G. B. Tangwa, In *A Companion to African Philosophy,* 387–395, by Kwasi Wiredu (ed.), 2004, Blackwell.

entails ethically informed planning that allows for coexistence and interdependence.

Taken together, these elements of African communitarian thought point toward a distinctive and deeply ethical approach to population and environmental sustainability. This approach recognizes the need to manage human population growth, not through coercion or fear, but through responsibility, justice, cultural sensitivity, and ecological awareness. It places the moral burden not on individuals alone, but on communities, traditions, and institutions that shape human behaviour.

This perspective affirms the importance of population control for environmental sustainability while grounding it in respect for human dignity and the wider web of life. It encourages dialogue among generations, between humans and nature, and across cultures, seeking not dominance, but balance. In this way, African communitarian ethics contributes meaningfully to global efforts to address the twin crises of population growth and environmental degradation.

8.6 Rethinking Population Control Through African Communitarian Ethics

In light of the preceding discussions, it becomes necessary to reconceptualize population control beyond the dominant Western technocratic or coercive frames that often prioritize demographic targets over moral and cultural contexts. An African communitarian ethical perspective offers a reorientation—one that views population ethics not merely as a matter of numbers, but as a question of how best to sustain flourishing human and ecological communities over time.

African communitarian ethics emphasizes the interconnectedness of individuals within the community, collective responsibility, cultural continuity, and the long-term survival of both human and nonhuman life. These principles support the idea that population control is ethically justifiable and even necessary, but only when pursued through non-coercive, inclusive, and culturally sensitive means that prioritize the common good and ecological harmony.

This rethinking draws from various African philosophical sources. The Akan idea of the *common good*, as articulated by Kwame Gyekye,[32] urges policies that promote shared well-being rather than isolated individual benefit. From this vantage, population policies must be judged by how well they enhance communal flourishing, equitable resource sharing, and intergenerational justice. Such policies must not impose external models, but instead emerge from within communities, with attention to local beliefs, values, and systems of support.

[32] From Person and community in African thought, by K. Gyekye, In *Person and community: Ghanaian philosophical studies I*, 101–122, by K. Wiredu & K. Gyekye (Eds.), 1992. Council for Research in Values and Philosophy.

Similarly, Godfrey Tangwa's[33] *live-and-let-live* ethics reinforces the imperative to control the human population, not as a means to diminish human worth, but to enable the coexistence of all life forms. Human expansion that depletes forests, displaces animals, and pollutes water sources violates this ethic. In contrast, responsible population practices guided by restraint and care uphold the moral imperative to live in balance with the wider ecology.

Importantly, African communitarian ethics rejects coercion. It affirms dignity and agency while emphasizing education, dialogue, and shared responsibility. Population control, in this view, is a communal endeavor rooted in moral persuasion and solidarity rather than enforcement. Programs such as community-based family planning, reproductive education, and support for women's empowerment align well with this ethic when pursued with cultural sensitivity and respect for human dignity.

8.7 Anticipating Critiques and Formulating Responses

In any complex ethical debate, particularly one as nuanced and contextually sensitive as population control in relation to African communitarianism, various critiques are likely to arise. The following are some of the most plausible objections to the approach outlined in this chapter, along with reasoned responses that emphasize the strengths and flexibility of the African communitarian perspective.

8.7.1 Critique 1: African Communitarianism Is Too Culturally Specific to Guide Contemporary Population Policy

Objection:

Critics may argue that African communitarian ethics, rooted in particular cultural contexts and oral traditions, may not provide a universally applicable or sufficiently modern framework for addressing global challenges like overpopulation and environmental degradation.

Response:

While African communitarianism arises from particular traditions, its core values, of interconnectedness, collective responsibility, sustainability, and respect for the common good, are not culturally exclusive. These resonate with broader human concerns and global sustainability ethics. Moreover, African ethical frameworks are dynamic and adaptable. Philosophers like Kwame Gyekye and Godfrey Tangwa

[33] From Some African Reflections on Biomedical and Environmental Ethics, by G. B. Tangwa, In *A Companion to African Philosophy,* 387–395, by Kwasi Wiredu (ed.), 2004, Blackwell.

have shown how African communitarianism can evolve to address contemporary issues. When interpreted in ways that engage modern realities, these values offer a robust foundation for ethical population policies that respect both cultural traditions and contemporary needs.

8.7.2 Critique 2: Emphasis on the Common Good May Override Individual Rights and Freedoms

Objection:

Western liberal critics might contend that communitarian approaches, by prioritizing the collective, risk suppressing individual autonomy, especially in areas such as reproductive rights.

Response:

African communitarianism does not negate individual dignity or rights; rather, it situates personal freedoms within a broader ethical context that emphasizes the moral duties individuals have toward others and their communities. The focus on collective responsibility is not a justification for coercion but rather for ensuring that personal actions, especially in areas like reproduction, do not harm the greater community or future generations. The ethics of *live-and-let-live* (Tangwa) and the long-standing commitment to justice and equity in African ethics seek to balance individual freedoms with communal well-being. Thus, ethical population policies can and should empower individuals while also promoting the common good.

8.7.3 Critique 3: Population Control Has a History of Abuse—How Does African Ethics Guard Against This? (8.7.3)

Objection:

Given the global history of coercive or discriminatory population control measures, including forced sterilizations and racially targeted policies, critics may question how any form of population control can be ethically justified.

Response:

This concern is valid and highlights the importance of guarding against coercion. African communitarian ethics provides safeguards precisely because it emphasizes human dignity, cultural sensitivity, and moral dialogue. It rejects top-down, externally imposed control measures in favor of participatory, community-led policies. The focus on shared ethical values, such as the right to dignity and collective responsibility, ensures that population control measures, when considered, are voluntary, respectful, and implemented with a focus on social equity. Ethical frameworks rooted in African communitarianism are grounded in respect for both individual autonomy and the collective well-being, which helps protect against abusive practices.

8.7.4 Critique 4: The Real Problem Is Consumption, Not Population

Objection:

Some environmental thinkers argue that the focus on population is misplaced. The problem lies more in overconsumption, particularly in affluent societies, than in population numbers per se.

Response:

This is an important and widely shared concern. Indeed, African communitarianism supports critiques of unjust consumption patterns and global inequality. However, in many African contexts, rapid population growth *does* place considerable strain on already scarce resources and fragile ecosystems. Population control measures, when implemented in an ethical and community-centred way, can help alleviate pressure on these resources. Moreover, focusing on population control in parallel with addressing overconsumption creates a more holistic approach to sustainability. Thus, while addressing consumption is critical, it should not overshadow the need to implement ethical and culturally sensitive population policies, particularly in regions facing rapid demographic changes.

8.7.5 Critique 5: African Traditions Value Large Families—Will Population Control Not Undermine This?

Objection:

Some may argue that traditional African societies view children as social and economic assets, and that, therefore, population control contradicts core cultural values that celebrate large families.

Response:

It is true that many African societies have historically celebrated large families. However, these values were shaped in contexts with different ecological conditions, economic systems, and mortality rates. In contemporary times, particularly with changing lifestyles, urbanization, and environmental degradation, the dynamics of family size and community support have evolved. African communitarian ethics does not require abandoning cultural values, but it does allow for their adaptation in response to modern challenges. Encouraging sustainable family planning through culturally grounded education and communal discussions can help families thrive in today's world without undermining tradition.

8.7.6 Critique 6: Who Decides What the "Common Good" Is?

Objection:

A practical concern arises about who gets to define the "common good" in a diverse and stratified society. Elites or political actors may misuse communitarian language to justify harmful or exclusionary policies.

Response:

This is a valid concern in any ethical or political framework. However, African communitarianism inherently values inclusivity and participatory decision-making. The "common good" is not a fixed, top-down concept; rather, it is arrived at through communal deliberation and consensus-building. In line with the principles of justice and equity, African communitarian ethics demands that decisions be made with the input of all community members, ensuring that policies respect diverse views and benefit everyone, particularly the marginalized. The idea of collective responsibility encourages the active engagement of all stakeholders in shaping policies that impact population and resource distribution.

8.7.7 Critique 7: Why Focus on African Ethics in a Global Issue Like Population Control?

Objection:

Some may question why African ethical frameworks should be foregrounded in what is a global environmental and demographic crisis, especially given the dominance of Euro-American perspectives in global discourse.

Response:

Population control and environmental sustainability are indeed global issues, but solutions must be locally grounded. Africa's demographic growth is one of the most significant globally, with implications for both regional and global sustainability. African communitarian ethics offers a unique lens through which to address these challenges, as it emphasizes collective responsibility, sustainability, and respect for future generations. While African traditions and ethical values may not offer a "one-size-fits-all" solution, they provide critical insights into how population policies can be framed in ways that respect local cultures and promote collective well-being. Including African ethical perspectives in the global conversation enhances the diversity of ethical thought and ensures that solutions are culturally relevant and contextually sensitive.

By addressing these critiques thoughtfully, we can ensure that the proposal for population control within African communitarian ethics is robust, nuanced, and sensitive to both local and global considerations. The goal is not to undermine individual freedoms or cultural values but to promote a balanced, ethical approach that ensures sustainability, justice, and dignity for all.

8.8 Conclusion

The debate over population control reveals a deeper tension in environmental ethics: how to balance the rights and dignity of individuals with the collective good of ecological sustainability. Utilitarian arguments often justify population control in terms of minimizing overall harm, while deontological and rights-based perspectives caution against treating people as means to an end. Social justice frameworks highlight the risks of reinforcing global inequalities, and environmental ethics stress the moral urgency of protecting the biosphere.

Ultimately, any viable approach must take seriously the moral complexity of these issues. It must resist simplistic solutions and recognize the importance of context, culture, and ethical nuance. In this regard, the dominant frameworks in Western thought, while valuable, are insufficient on their own. What is needed is a broader ethical conversation that includes alternative paradigms, particularly those rooted in relational and communitarian philosophies.

This chapter has explored how traditional African communitarian ethics provide such a paradigm, offering an alternative ethical lens through which to understand and respond to the intertwined challenges of population and environmental sustainability. This chapter has argued that rethinking population ethics and environmental sustainability through African communitarian values offers a much-needed ethical lens to address the twin challenges of population growth and ecological degradation. Rather than imposing abstract or coercive frameworks, African traditions offer a rich ethical soil for cultivating policies and practices rooted in relational responsibility, interdependence, and respect for future generations.

The communitarian ethic, which draws from Ubuntu, the concept of the common good in Akan thought, and Tangwa's principle of *live-and-let-live*, emphasizes that living well requires us to live together—humans, nonhumans, and ecosystems alike. Ethical approaches to population must therefore extend beyond merely counting bodies; they should focus on nurturing the kinds of communities that sustain life in all its diversity. Ultimately, a truly sustainable future must be built on ethics that recognize both the limits of the Earth and the obligations of care, solidarity, and justice. African communitarian philosophy offers not only a critique of current paradigms but also a hopeful path forward: one that affirms human dignity while securing the conditions for all forms of life to flourish.

References

Bakare, K., and S. Gentz. 2020. Experiences of forced sterilisation and coercion to sterilise among women living with HIV (WLHIV) in Namibia: An analysis of the psychological and socio-cultural effects. *Sexual and Reproductive Health Matters* 28 (1): 1758439. https://doi.org/10.1080/26410397.2020.1758439.

Behrens, K. G. 2012. Moral obligations towards future generations in African thought. *Journal of Global Ethics* 8 (2–3): 179–191. https://doi.org/10.1080/17449626.2012.705786.

Broome, J. 1992. *Counting the cost of global warming*. Cambridge: White Horse Press.
Cohen, J. E. 2010. The earth is fast approaching a population limit. Managing global population growth is critical for sustainability. *Population and Development Review* 36 (4): 645–655.
Connelly, M. 2008. *Fatal misconception: The struggle to control world population*. Harvard University Press.
D'Alessandro, S., et al. 2017. Population stabilisation and climate change mitigation: A global perspective. *Environmental Science & Policy* 77:1–10.
Ehrlich, P. R. 1968. *The population bomb*. Ballantine Books.
Ehrlich, P. R., and J. P. Holdren. 1971. Impact of population growth. *Science* 171 (3967): 1212–1217.
Eze, M. O. 2008. What is African communitarianism? Against consensus as a regulative ideal. *South African Journal of Philosophy* 27 (4): 386–399.
Gyekye, K. 1992. Person and community in African thought. In *Person and community: Ghanaian philosophical studies I*, ed. K. Wiredu and K. Gyekye, 101–122. Council for Research in Values and Philosophy.
Gyekye, K. 1997. *Tradition and modernity: Philosophical reflections on the African experience*. Oxford University Press.
Hartmann, B. 1995. *Reproductive rights and wrongs: The global politics of population control*. South End Press.
Hoffert, M. I., et al. 2002. Advanced technology paths to global climate stability: Energy for a greenhouse planet. *Science* 298 (5595): 981–987.
Kant, I. 1785. *Groundwork for the metaphysics of morals*. Cambridge University Press.
Leopold, A. 1949. *A Sand County Almanac*. Oxford University Press.
Malthus, T. R. 1798. *An essay on the principle of population*. London: J. Johnson.
Martine, G., and D. Schensul. 2013. *The demography of adaptation to climate change*. UNFPA, IIED.
Mbiti, J. S. 1990. *African religions and philosophy*. Heinemann.
McMahan, J. 1981. Problems of population theory. *Ethics* 92 (1): 96–127.
Narveson, J. 1973. Moral problems of population. *The Monist* 57:62–86.
Parfit, D. 1984. *Reasons and persons*. Oxford University Press.
Peters, G. P., et al. 2011. Growth in emission transfers via international trade from 1990 to 2008. *Proceedings of the National Academy of Sciences* 108 (21): 8903–8908.
Rawls, J. 1971. *A theory of justice*. Cambridge, MA: Harvard University Press.
Silliman, J., and Y. King, eds. 1999. *Dangerous intersections: Feminist perspectives on population, environment, and development*. South End Press.
Tangwa, G. B. 2004. Some African reflections on biomedical and environmental ethics. In *A companion to African philosophy*, ed. Kwasi Wiredu, 387–395. Blackwell.
Tutu, D. 1999. *No future without forgiveness*. New York: Doubleday.
United Nations. 2022a. *World population prospects 2022*. https://population.un.org/wpp/
United Nations. 2022b. *World population prospects 2022. Summary of results*. United Nations Population Division.
Wiredu, K. 1996. *Cultural universals and particulars: An African perspective*. Indiana University Press.
World Bank. 2021. *Lagos urban transport and environmental assessment report*. World Bank.

Chapter 9
An African Population Ethics and the Environment: Addressing the Impact of Global Warming on the Well-Being of African Peoples

Keba Cyril Muko

Abstract Well-being is a multidimensional reality that must take into account the different aspects of the individual, collective, and social life of human beings. What constitutes well-being has many facets, such as feeling good, having a balanced life, being under a governance system that enables people to achieve their goals, and having the means to do so. One must envision a holistic policy of well-being. In this policy, the individual, as a conscious being, is called to realize the self by the act of openness to the other, who is also a conscious subject, and equally, openness of the self to the other entity that is non-rational but substantially important, the environment. In this way, the individual's reflexive self-consciousness interlocked with the relational subjectivity of the other, mutual intersubjectivity can reclaim the African sacredness of the environment with which populations have a rationalized and humanized encounter for their well-being. It is, therefore, an ethical perspective on reflexive self-consciousness and intersubjectivity that must strengthen the creation of a population attuned to the existential openness to the environment if well-being is to be attained. Such a scheme of well-being is to be conceptualized by asking the question: how can the creation of a population be attuned to the reflexive self-consciousness and African intersubjectivity in order to reclaim the existential aperture to the environment to mitigate the negative impacts of global warming that affect the well-being of African populations? In an approach which adopts African ethical analysis and objective argumentation, I intend to generate a rational narrative that contributes to the knowledge of the issues of population ethics, especially thematized on the African ethics of intersubjectivity with regard to the environment as a shared good for the well-being of African populations, present and future.

Keywords Self-consciousness · African intersubjectivity · Environment · Populations · Ethics · Person-based-theory

K. C. Muko (✉)
Faculty of Philosophy, St. Bonaventure University, Lusaka, Zambia
e-mail: dean.office@sbu.ac

B. Okyere-Manu, P. Mwipikeni (eds.), *An African Approach to Population Ethics*, Advancing Global Bioethics 23,
https://doi.org/10.1007/978-3-032-24464-2_9

9.1 Introduction

In this study, we aim to design a population ethics based on the interpersonal ethical point of view which must stir up a self-consciousness such that one grasps that the "acceptable" act is that which knows that if the well-being of present and future populations is to be thinkable and achievable, the African person must endorse the environment as a relational subject with which one must collaborate and communicate.

The standpoint of this design is African intersubjective ethics which takes into account the fact that humanity in an individual is successfully achieved only when the humanity in the other individual is equally recognized as such, thus focusing on community rather than individuality.[1] Such intersubjectivity must include the act of being proper to the environment. It is in this idea of communion that the mutual well-being of interconnected and relational peoples is conceived, both the existing African populations and those still to be called into existence. The focus here is less on happiness, considering that the concept of well-being is more encompassing and more fundamental, as it includes happiness, which, in turn, can increase well-being. Clearly, at the level of general ethics, the concept of happiness has occupied a preponderant place in the thought of philosophers such as Plato, Aristotle, John Locke, Thomas Jefferson, and many others. Aristotle, being a pioneer of ethical reflection on happiness, affirmed that "he is happy who lives in accordance with complete virtue and is sufficiently equipped with external goods, not for some chance period but throughout a complete life."[2] In this sense, Aristotle considered happiness as something to be pursued for a lifetime, and which cannot be said to have been attained until the end of a person's life. Happiness is not a temporary state; rather, it is a goal that is to be pursued. Understood in this perspective, happiness can be encompassed in the concept of well-being, which is equally a goal more than a temporary state, though it entails a sense of temporality and historicity as a person's life span is within temporality and historicity.[3] The concept of well-being in philosophy can mean all that is in itself good for a person. These are the intrinsic or basic welfare goods and not anything else that may be instrumentally good in the sense of a derivative good. Thus, well-being is that which is intrinsically good for the person.[4]

This point of departure is intended to lay a foundation for a reflection that is built on the African ethics of intersubjectivity, which equally clears the way to the

[1] Bénézet Bujo, "Ecology and Ethical Responsibility from an African Perspective," in African Ethics: An Anthology of Comparative and Applied Ethics, ed. Munyaradzi Felix Murove (Pietermaritzburg: University of KwaZulu-Natal Press, 2009), 281–97.

[2] Aristotle, Nicomachean Ethics, trans. Terence Irwin (Indianapolis: Hackett, 1999), 1101a14–16.

[3] Heidegger, M., 2002. *On time and being*. University of Chicago Press; Hoy, D.C., 2012. *The time of our lives: A critical history of temporality*. MIT Press.

[4] Meissner Darius, 2021. Population Ethics and Theories of Well-Being. [Online]. Available from: www.utilitarianism.net/population-ethics. [Accessed 2 September 2023].

understanding of well-being, a state of life achieved through adherence to the ethical, spiritual, and communal social life of the person. African morality goes hand in hand with all that gives meaning to life, nourishes it, maintains it, and makes it better and better; and this is the fulfillment of the African human nature through becoming an ethical person.[5] There are rituals, traditions, and ceremonies to cement daily life in perfect harmony between the living and the dead, spirits and other beings of nature. Choosing good, doing good, not doing harm or causing harm, is the moral code that emerges from the lived experience, and that is why it is right to admit that African morality is experienced and not codified in metaphysical principles; we experience it in daily life.[6] In this context, we understand that for a population ethics constructed around self-consciousness and the African approach of intersubjectivity, the African notion of ethics is fundamental because the African person understands himself better as a self-determined and self-realized subject, in the relational reciprocity of subjects; in short, in the affirmation of intersubjectivity.[7] This reciprocity of subjectivity includes a fundamental relationship to the environment in all its diversity and vital systems. My approach to population ethics is built on this African vision of ethics, of which the dimension of African intersubjectivity is foundational.

Accordingly, it should be admitted that population ethics has, besides the wellfare of existing populations, a direct link with the consideration and far-sightedness for the happiness of people to come in the future.[8] As far as the environment is concerned, we adopt the same type of far-sightedness and anticipation of welfare for existing and future populations. When we act in relation to the environment, we must necessarily deliberate about not only the welfare of the existing populations but also of the populations that we have the responsibility to bring into this world. Thus, the concept of "welfarist principles" as debated by Charles Blackorby and others[9] will guide our reflection. Welfarist principles can be comprehended as indicators of a lifetime well-being, not of a single individual, but of all individuals.[10] Hence, our task is the bridging up of self-consciousness and intersubjectivity in an existential reciprocity of subjects, intelligible and non-intelligible in the being of the environment. Such connectivity, manifested as concern for the welfare of existing and future populations, ought to be a lived and verifiable reality, reinforced in an

[5] Shutte, A., 2008, "African Ethics in a Globalizing World." In: *Persons in Community. African Ethics in a Global Culture*, Ronald Nicolson (Ed.) University of Kwazulu Natal Press, 15–34. 30.

[6] Mkhize, N., 2008. Ubuntu and harmony: An African approach to morality and ethics. In: *Persons in Community. African Ethics in a Global Culture*, Ronald Nicolson (Ed.) University of Kwazulu Natal Press, 35–44.

[7] Shutte, African Ethics in a Globalizing World. 26–27.

[8] Huseby, R., 2012. Sufficiency and population ethics. Ethical Perspectives, 19(2), pp. 187–206.

[9] Blackorby, C., Bossert, W. and Donaldson, D., 2004. Critical-level population principles and the repugnant conclusion. In *The repugnant conclusion: Essays on population ethics* (pp. 45–59). Dordrecht: Springer Netherlands.

[10] Ryberg, J. and Tännsjö, T. (eds), 2004. *The repugnant conclusion: Essays on population ethics*. Springer Netherlands.

approach that can be read through certain indicators of well-being. The same indicators that determine the threshold of well-being should be considered when illustrating African population ethics. One of the indicators is the omnipresent question of a healthy and suitable environment and how existing populations, taking into account the needs of future populations, act to alleviate or accelerate initiatives toward climate solutions in the presence of natural disasters that are becoming more and more deadly. Self-consciousness and intersubjectivity are two significant concepts that have evolved in recent years, even though numerous other aspects also affect a person's overall sense of well-being.[11] Self-consciousness is the awareness of a person's thoughts, feelings, and self-constructed and transmitted experiences,[12] whereas intersubjectivity is the capacity of people to comprehend and interact with one another.[13] Both of these concepts are crucial to one's general well-being and are intimately related to how one interacts with their surroundings, the environment in particular. In this study I will examine the connections between intersubjectivity, self-consciousness, and the environment and make the case that these concepts are crucial to a person's overall well-being.

First, I will explain the ideas of self-consciousness and African intersubjectivity as a fundamental moment of personal and interpersonal understanding and self-disclosure, with an important place accorded to the understanding of the environment, without which there is no attainment of individual or collective well-being. To emphasize the link between the environment and well-being, I reflectively bring forth a narrative of the impacts of global warming and the existential vulnerability of the African populations concerned. The point is that the harmful impacts of the climatic crisis accelerate the existential vulnerability of the existing African populations and already compromise the foresight of future populations. These negative impacts are a major reason to begin an ethical evaluation of human actions, as we know there is no population ethics without this evaluative moment. This may compel the existing population to reclaim the sense of the sanctity of the environment in visualizing the transfer of the same sense of the sacredness to future populations. With a vision turned toward future populations, there is a thought-out pause to mark the reality of the birth rate in Africa.

Understanding African intersubjectivity as an existential mode, we quickly realize that the African person's way of being and way of moral actions are intertwined to the extent that, according to African culture and sociability, a person is only validated when his or her actions are carried out in a way that builds cohesion with other beings affected by the actions taken.[14] As a result, the collective expectation is that

[11] Gillett, G. and McMillan, J., 2001. *Consciousness and intentionality* (Vol. 27). John Benjamin Publishing.

[12] Wallace, B. and Kunzendorf, R.G., 2000. Individual differences in conscious experience. *Individual Differences in Conscious Experience*, pp. 1–424.

[13] Husserl, E., 2014. *Ideas for a pure phenomenology and phenomenological philosophy: First book: General introduction to pure phenomenology*. Hackett Publishing.

[14] Husserl, E., 2014. *Ideas for a pure phenomenology and phenomenological philosophy: First book: General introduction to pure phenomenology*. Hackett Publishing.

the African existential mode, based on intersubjectivity and self-consciousness, replicates a person's moral character. To this extent, African ethics arises from this interconnectedness, and, in my view, this gives rise to the ethics of intersubjectivity necessary.

9.2 Interlocking Self-Consciousness and the African Ethics of Intersubjectivity

From the onset, I would like to clarify the African concept of intersubjectivity, within the framework of African ethics, which I thematize as the African ethics of intersubjectivity; then I also resort to the philosophical notion of the concept of intersubjectivity. I find it essential to link the two notions of the intersubjectivity understood in the context of African ethics and also in the perspective of philosophy in general. Indeed, African ethics is defined in line with the sociability of the African people, of their societal customs in relation to moral actions and their meaning in terms of traditions, rituals, and ceremonies that mark a sense of communal identity.[15] In this context, an African person and individuals understand themselves within a precise structure of the community, because it is by the openness to other individuals that the African person has a better and clearer awareness of his/her being.[16] This amounts to saying that to know oneself better, the person must open up to others and receive feedback from others; moral actions constitute an instance of self-consciousness of the person because the African person is able not only to carry his own being but also to reflect on his own actions. The person must first realize himself as a subjective being before better understanding himself or herself as an intersubjective being. African intersubjectivity is therefore understood within the framework of the community of individuals conscious of their own being, adhering to the vital model of communal African solidarity, and observing the rules with the aim of increasing vital force.

I emphasize here that the notion of African intersubjectivity differs from the individualistic notion of life specific to other cultures. In the context of African cultures still detached from the strong modern influences of foreign cultures, intersubjectivity is a way of being human forged by the networks of community relations, shared values, experiences, and histories of peoples and culminating in a

[15] Mafunisa, J.M., 2008. Ethics, African societal values and the workplace. In: *Persons in Community. African Ethics in a Global Culture*, Ronald Nicolson (Ed.) University of Kwazulu Natal Press, 111–124, 112; Richardson, N., 2009. "Can Christian Ethics find its way, and itself, in Africa?" In Murove, M.F (Ed.) *African Ethics. An Anthology of Comparative and Applied Ethics*. Pietermaritzburg, University of Kwazulu-Natal Press, 129–154. 131.

[16] Munyaka, M. and Motlhabi, M., 2009. "Ubuntu and its socio-moral significance." In Murove, M.F. (Ed.) *African Ethics. An Anthology of Comparative and Applied Ethics*. Pietermaritzburg, University of Kwazulu-Natal Press, 63–84. 68.

connectedness that engenders a certain personal and collective identity.[17] African intersubjectivity denotes the importance of interpersonal relationships in harmony with spirits and the environment, and also manifests a very dynamic social organization among groups recognizing themselves in a certain specific culture.[18] This is the conception of intersubjectivity which generates a mode of interpersonal conduct essentially mindful of the other person, be it human or non-human, alive or dead. This type of ethics, an African ethics of intersubjectivity, favors the communal sense of "We", not by absorbing the "I", but by activating a communion and harmony between the different "I's" of society.[19] The African ethics of intersubjectivity is a harmonious and always relational openness to all other beings that can contribute to the life force.

Knowing oneself and the other, as a rational subject, in a "I–thou; I–it" relation with another human person,[20] and with a non-rational subject, the environment is an essential indicator and an overture to well-being. This is because, to have the consciousness of oneself as a singular, particular being distinct from the other in the depths of the very being of oneself, and to become imbued with the said consciousness and its knowledge, is important if the individual is to access self-well-being. Most Africanists hold the view that the African person is a unity of the spiritual and material aspects of life. The physical world in all its visible aspects is only a continuation of the spiritual world, which encompasses all the elements that sustain the harmony of the vital force. From the moment the person in his or her daily actions knows how to maintain the physical and the material worlds in symbiosis, this creates psychological and social balance; thus, the person becomes total in the being of self. The person will feel in deep balance in their own being, and also in practical balance with others, thus constituting a solidarist society and community, a source of an admirable level of well-being.[21] This means that the person feels well in her/his physical, psychological, and spiritual being, and this can also extend to the social and economic wellness in accordance with the enablers put in place by those

[17] Mafunisa, J.M., 2008. Ethics, African societal values and the workplace. In: *Persons in Community. African Ethics in a Global Culture*, Ronald Nicolson (Ed.) University of Kwazulu Natal Press, 111–124.

[18] Mkhize, N., 2008. Ubuntu and harmony: An African approach to morality and ethics. In: *Persons in Community. African Ethics in a Global Culture*, Ronald Nicolson (Ed.) University of Kwazulu Natal Press, 35–44.

[19] Dube, M.W., 2009. 'I am because we are': giving primacy to African indigenous values in HIV and AIDS prevention. In Murove, M.F (Ed.) *African Ethics. An Anthology of Comparative and Applied Ethics*. Pietermaritzburg, University of Kwazulu-Natal Press, 221–237.

[20] Buber, M., 1970. *I and Thou* (Vol. 243). Simon and Schuster.

[21] Bujo, B., 2009. Ecology and ethical responsibility from an African perspective. In Murove, M.F (Ed.) *African Ethics. An Anthology of Comparative and Applied Ethics*. Pietermaritzburg, University of Kwazulu-Natal Press, 281–297; Abraham, W.E., 2015. The mind of Africa. African Books Collective; Wiredu, K and Gyekye, K., 1992. Person and community: Ghanaian philosophical studies I (Vol. 1). Cipsh/Unesco. The Center for Research in Values and Philosophy.

with such a mandate.[22] Feeling good about oneself can also mean having good living conditions in terms of economic outcomes. We should further add that feeling good in one's being also means the emotional amount of joy felt in inter-human relationships and in everything that contributes to increasing life, for example, an environment favorable to life.

For the most part, it is therefore not possible to separate true self-consciousness and the true well-being of oneself and others. However, it is also necessary to better define what is meant by self-consciousness. There are a variety of ways to understand and explain self-consciousness. Different philosophers contributed their thoughts on the topic. Among others, Piaget, Merleau-Ponty, and Rochat.[23] In their thoughts, a consistent theoretical model was established by Piaget, insisting on an innate self-consciousness, that the person is born as a subject, knowing that he/she is a subject at least in his/her bodily dimension. Building on Piaget's developmental psychology, Merleau-Ponty focused on the phenomenology of perception. Recalling the idea of innate self-consciousness, he emphasized the important dimension of the other for individual self-awareness. Further thinkers have argued that for an individual, self-consciousness is the awareness of a person's own body as located in a time-space frame while interacting with others, including the environment.[24] In this sense, self-consciousness also comprises the awareness of the person's own identity, which is built within time and space through interactions with others and the environment. Consequently, self-awareness becomes a process of identification and of differentiation of the self and others in their image and emotional experiences.[25]

Thus explained, self-consciousness can mean the person's capacity for introspection while becoming more aware of one's existence and distinctiveness as a unique being from others, while equally being conscious of the uniqueness of others. Being conscious of oneself is essential to general well-being since it enables people to comprehend and process their feelings and their very existence, and all that it entails. Self-aware people are gifted with the capacity to be fully in touch with their surroundings and to regulate their behavior and engage in choices that are consistent with their objectives and ideals in order to access well-being.

Self-consciousness is also well understood when it is perceived together with intersubjectivity. This is the notion that, in alignment with the views of Piaget and Merleau-Ponty of an innate self-presence, it has also been defended by social constructivism, that "the self is constructed and negotiated through social interaction

[22] ACET, 2023. Human well-being in the African Transformation Index. [no date]. Available at https://acetforafrica.org/ati/growth-with-depth/human-well-being/. Accessed 25th March 2024.

[23] Piaget, J. (1936). The Birth of Intelligence in Children. Neuchatel: Delachaux and Niestlé; Merleau-Ponty, M. (1964). Relationships with others in children. Bull. Psychol. 18, 295–336.

[24] Decety, J. and Sommerville, J.A., 2003. Shared representations between self and other: a social cognitive neuroscience view. *Trends in cognitive sciences*, 7(12), pp. 527–533.

[25] Decety, J. and Sommerville, J.A., 2003. Shared representations between self and other: a social cognitive neuroscience view. *Trends in cognitive sciences*, 7(12), pp. 527–533.

and that all self-experience is intersubjectively mediated."[26] The idea of intersubjectivity as an indicator of well-being holds that social interactions and communication between people include the awareness of others' world of consciousness, and a comprehension of each other's experiential selves and views as in an emphatic reciprocal understanding.[27] Social interactions lead to a shared understanding and interpretation of events that are meaningful and constructive to the selves. In the presence of each other, individuals admit that social and cultural exchanges magnify intersubjectivity. Intrinsically, social and cultural exchanges have an impact on individuals' subjective experiences, and their interactions with others have an effect on how they perceive and understand the world in front of them, the public world, a world in which they are constructing their well-being, individually and conjointly.[28] Intersubjectivity, which enables people to interact with others and develop lasting relationships, is a decisive indicator in overall well-being, as it is an openness from the Cartesian essence of selfhood understood as a self-enclosed, solitary interiority.[29] that is close to self. When intersubjectivity is taken seriously, it must lead the person to understand that the right and correct action toward others is that which, in one way or another, respects their humanity or their very essence, and that determines one's conduct toward them. Accordingly, a proper appropriation of the notion of African intersubjectivity must necessarily lead to an ethics of intersubjectivity.

The point we make here is to emphasize that it is this ethics of intersubjectivity that we grasp as the founding theory of our Approach to an African Population's Ethics. More clearly, such African Populations ethics is based on the following maxim: Act in such a way that the impacts of your actions on humans and on the environment are always oriented toward achieving the well-being sought by present populations, while at the same time, postulating the qualitative well-being of populations still to be called into existence. Thus, one cannot comprehend the maxim of African Populations' ethics without adopting the mode of being proper to African intersubjectivity that is intertwined with its mode of conduct or its ethics. To this end, we can appeal to the sense of *Ukama ethics*, based on a Shona word from Zimbabwe, which reminds us of our common humanity and our close relationship with nature. According to Professor Murove, "an ethic of Ukama is an ethic of holism. It is holistic in the sense that my well-being as an individual has its

[26] Zahavi, D., 2014. *Self and other: Exploring subjectivity, empathy, and shame*. Oxford University Press, USA.

[27] Branco, A.U. and Lopes-de-Oliveira, M.C. eds., 2017. Alterity, values, and socialization: Human development within educational contexts (Vol. 6). Springer.

[28] Keromnes, G., Chokron, S., Celume, M.P., Berthoz, A., Botbol, M., Canitano, R., Du Boisgueheneuc, F., Jaafari, N., Lavenne-Collot, N., Martin, B. and Motillon, T., 2019. Exploring self-consciousness from self-and other-image recognition in the mirror: concepts and evaluation. *Frontiers in psychology*, Volume 10.

[29] Zahavi, D., 2014. *Self and other: Exploring subjectivity, empathy, and shame*. Oxford University Press, USA.

ontological meaning in relationship to my fellow human beings, those who are still alive, ancestors, God, and the environment."[30]

None can achieve well-being all alone; well-being in isolation is no well-being at all. African intersubjectivity is a chance to exit the shell of the self in an embrace of the humanity of other Beings. Well-being can only be considered in a context of a population, seen not as closing in on itself, but as a population that understands itself in openness to others. In the African context, this is more vivid and expressive. Individuals are better off when they develop trust and cooperation, which can result in stronger social and cultural interactions.[31] Intersubjectivity, like self-consciousness, can be obstructed by a lack of trust and cooperation due to prejudices, social exclusion, and cultural absolutism. People must be able to interact with others meaningfully and authentically in order to bridge self-consciousness and intersubjectivity, which can be challenging in situations when interaction is hindered by linguistic or cultural barriers. From now on, there is a need for a new approach regarding the African populations that we must consider bringing into existence. In a word, such a population tending towards the ascent to well-being must be a population that comes into the world in family and societal structures and institutions well prepared for the existential spaces of coordination between self-consciousness and intersubjectivity. This is all the more important to the extent that the notion of intersubjectivity is a fundamental notion in the conception of the human person in Africa. The human person's value in Africa, borne out of shared humanity, entails an intersubjective human being in an equally intersubjective community to the extent that the growth of the individual is directly linked to the advancement of the community.[32] This would mean that the well-being of the person is closely associated with the well-being of the community. Individuals are open to each other and to the community of persons. It is this type of an intersubjective population that we envisage. This is a population that would enter into a society where education and instruction are already part of the logic of the permanent openness of subjects to each other. In this openness of the subjects, there is another opening which must be undertaken; it is that of the intelligent subjects, toward the non-intelligent beings, in particular, nature and the environment, in which, without this man–nature intersubjectivity, meaning the interaction with the environment, any possibility of well-being can only remain an impossible possibility. This is especially true since in the absence of a rational and normative interaction with the environment, we can, unfortunately, only experience catastrophes that we could have avoided.

---

[30] Murove, M.F., 1999. *The Shona concept of Ukama and the process philosophical concept of relatedness, with special reference to the ethical implications of contemporary neo-liberal economic practices* (Doctoral dissertation, University of Natal, Pietermaritzburg). 1.

[31] Nicolson, R. (Ed.), 2008. Persons in community: African ethics in a global culture; University of Kwazulu Natal.

[32] Shutte, African Ethics in a Globalizing World. 31.

9.3 A Narrative of the Impacts of Global Warming and the Existential Vulnerability of the African Populations Concerned

The African person has always lived in harmony with the cosmos because in his/her deep thinking and in his/her actions, he/she has always understood that to live better, he/she had to maintain correlation and co-existence with all the other important beings in his/her immediate surroundings, whether it was plants, animals, rivers, or mountains. More importantly, the African person conformed to a sort of code of morality in his relationships to these beings of the environment.[33] This amounts to emphasizing the fact that Africans, conscious of traditions, were less ignorant of the responsibility towards ecology.

In contrast, such awareness seems to have waned in the thoughts and actions of many. For this reason, there is now a great call for the return of such awareness in relation to the responsibility of humans toward ecology. The existential balance between human well-being and the safeguarding of the environment is essentially becoming a renewed vision of the modern person, which has just awakened after several years of sometimes unnoticed or sometimes deliberate neglect by populations. It is a renewed vision because the traditional African person was well aware of such an existential symbiosis between nature and humans and consented to observe and respect it. The traditional man respected the environment without being constrained by external factors because he/she was fully aware that his/her well-being largely depended on that interdependence. Such a notion, over time, lost by the moderns, finds itself timidly renewed in the consciousness and experiences of people.[34] Accordingly, at this turn of the history of humanity, the population is invited to become imbued again with the extreme importance of rebuilding and maintaining the existential balance between happiness, well-being, and the safeguarding of the environment.

In recent years, as a result of the refusal to recognize the need to maintain the existential balance of the per-equation human well-being equals the well-being of the environment, we are witnessing real natural disasters that do not need supernatural explanations, only scientific demonstrations of the negligence of the said per-equation. We rush to manage the risks of natural disasters, instead of recommending preventive measures upstream.[35] Far from us, the deliberate idea of rejecting these

[33] Bujo, B., 2009. Ecology and ethical responsibility from an African perspective. In Murove, M.F (Ed.) *African Ethics. An Anthology of Comparative and Applied Ethics*. Pietermaritzburg, University of Kwazulu-Natal Press, 281–297; Abraham, W.E., 2015. The mind of Africa. African Books Collective; Wiredu, K and Gyekye, K., 1992. Person and community: Ghanaian philosophical studies I (Vol. 1). Cipsh/Unesco. The Center for Research in Values and Philosophy. 281–286.

[34] Dietz, T., Millar, D., Dittoh, S., Obeng, F. and Ofori-Sarpong, E., 2004. Climate and livelihood change in North East Ghana. In *The Impact of Climate Change on Drylands: With a Focus on West Africa* (pp. 149–172). Dordrecht: Springer Netherlands.

[35] Hazell, P., 2004. Climate change and management of catastrophic risk. In *The Impact of Climate Change on Drylands: With a Focus on West Africa* (pp. 385–395). Dordrecht: Springer Netherlands.

risk management measures, we intend here to advocate, at the very level of population management, new dynamics of present and future populations that sufficiently include the repugnant conclusion adapted to our repugnant question. Should we first seek the well-being of the population, or should we first be reassured of the well-being of the environment? Should these two be separated? A quick response is found in the search for an existential balance.

By existential balance, it can be understood that, on the one hand, as human populations must maintain themselves alive in a world of technological advancements, normally, well-being should include the socio-economic dimensions, with whatever that may imply. On the other hand, for progress to be made in the economic field, human societies must act on nature and the environment, which already implies a sort of "violence" done to the environment, because, for example, new economic infrastructures must be built in places where there may have been a tourist grove. However, the emergence of economic progress cannot necessarily imply excessive violation of the environment. Certainly, from the point of view of the co-existence of both (humans and the environment), even if socio-economic advances mean necessary actions on the environment, it is recommended that the normative imperative at all costs for the balance of actions; otherwise, there is a risk of falling into the extremism of economic overexploitation or into a sort of environmental enslavement. On this aspect, it was observed that "the greatest current efforts to 'save' Africa are associated with contemporary environmentalism" [because] Under the banner of saving the African environment, Africans in the last half century have been subjected to a new form of "environmental colonialism."[36] If human populations do not consider the quality of the environment as relevant to the quality of their own lives, and if socio-economic progress ignores the merits of the quality of the environment, human populations will eventually perish. An existential balance is necessary, and we had better be aware of it.

Several African countries have faced countless disasters of magnitudes never experienced before. For example, as reported by Africa news, "between 8 and 21 April 2022, the slow-moving storm Issa brought long periods of heavy rain to KwaZulu-Natal, which caused flooding and mudslides in Durban and surrounding areas, affecting over 40,000 people and leaving a trail of destruction."[37] Another disaster extended its rays of destruction, traveling from Madagascar to Mozambique via Malawi. A report by Al Jazeera indicated that "the death toll from Tropical Cyclone Freddy has passed 300 as the body count continues to rise, with authorities in Mozambique and Malawi taking several days to assess the extent of the damage and loss of life. The storm tore through Southern Africa over the weekend for a second time after first making landfall in late February. It is one of the longest-lasting tropical cyclones ever recorded, and one of the deadliest in Africa in recent

[36] Nelson, R.H., 2003. Environmental colonialism: "saving" Africa from Africans. *The Independent Review*, *8*(1), pp. 65–86.

[37] Africa News, "South Africa: KwaZulu-Natal Floods," Knowledge Hub, April 8, 2022, accessed November 4, 2025, https://www.jbarisk.com/knowledge-hub/event-response/south-africa-kwazulu-natal-floods/

years. Malawi has reported 225 dead so far, with hundreds more injured and some still missing. The storm had killed about 27 people in Madagascar and Mozambique before lashing Mozambique for a second time."[38] Recently, the eastern part of the DRC woke up discovering that the entire compound and village of Kalele had been washed away as narrated by France 24: "the death toll from floods and landslides in the east of the Democratic Republic of Congo has risen to nearly 400, an administrative official said Sunday."[39]

Even in the absence of endorsed scientific evidence, we would not hesitate to attribute these named disasters to climate change caused by human actions, whose impacts prove to be callous on the same humans whose ascent toward well-being for some is fatally destroyed, while for others, well-being is irreparably compromised.

The effects of climate change and global warming on human well-being are often irreversible. Human well-being, severely impacted by catastrophic climatic change and global warming, is intergenerational. Yet, responsibilities vary in terms of which generations bear the most responsibility and which generation has discarded the obligation of generational equity in terms of accelerating the imminence of catastrophic climatic change.[40] When a generation behaves in an irresponsible way by creating the conditions for the destabilization of the climatic balance, such a generation compromises its own well-being; at the same time, it compromises the equitable distribution of well-being with regard to future generations. Such behavior is legally and ethically blamable insofar as the actions and decisions of the present generation dictate and decide the way of life of the generations of the future. The dilemma is that in their decisions and actions, seeking to increase the possibilities for their betterment and well-being, the current African populations may, in the process, create situations that may later claim possibilities of well-being for future populations. Hence, one can wonder if any choice to increase possibilities of well-being for a given population is really a rational prioritization, insofar as it seems to ignore the well-being of individuals of other generations to come.[41] Such a philosophical wonder necessarily leads to an ethical evaluative perspective of the actions of the existing populations and puts them face to face with the concomitant responsibilities towards the populations to come.

---

[38] (Aljazeera News, 16 March 2023).

[39] (France 24, 2023).

[40] Méjean, A., Pottier, A., Fleurbaey, M. and Zuber, S., 2020. Catastrophic climate change, population ethics and intergenerational equity. *Climatic Change*, *163*(2), pp. 873–890.

[41] Méjean, A., Pottier, A., Fleurbaey, M. and Zuber, S., 2020. Catastrophic climate change, population ethics and intergenerational equity. *Climatic Change*, *163*(2), pp. 873–890.

9.4 Ethical Evaluation of Human Actions on the Environment

In the past few years, the world's population has increased exponentially, reaching 8 billion inhabitants, thereby raising attention on ways and means to care for the populations in terms of food and other human needs, placing a huge strain on the environment. On the other hand, the world's population is becoming more aware of the pressing necessity for environmental conservation and protection. Therefore, it is extremely important to examine the actions of individuals and societies toward the environment. In particular, evaluating the actions of the African human persons toward the environment. For centuries, the African populations have relied excessively on natural resources for their livelihood and survival, and at the same time, there is a necessity to recognize the connectedness of African populations to the environment at all levels, based on African tradition and cultures.[42] African populations have a unique relationship with the environment, given the continent's rich natural resources and diverse ecological systems. However, as modernization and development continue to shape the continent, it is crucial to assess the moral implications of the actions taken toward the environment and provide a comprehensive analysis of the ethical considerations that must be taken into account. We must, however, remember that, in an anthropocentric vision, human actions on nature are very often justified by the supreme good that humans have in using nature to keep themselves alive, especially since they believe that nature is a good for the ultimate utility of humans.

Morally, are humans justified in thinking this way? "According to person-based consequentialism, or PBC, whether a particular future X is morally permissible or correct, depends on and only on whether existing and future persons at X are or will be treated in a morally permissible or correct manner."[43] The ethical theory of consequentialism states that an action is judged morally right or wrong based on its consequences over the people concerned; this is also known as total utilitarianism which determines the rightness or wrongness of an action based on aggregate utility it brings to the majority of those involved—in population ethics, a new way of considering the consequentialist theory is known as person-based or person affecting consequentialism; a person-based consequentialism states that an act is bad only if it is bad for somebody existing or for somebody in the future; further, an act is good only if it is good for someone existing or for some future population—PBC conveys agents to act in a way that persons affected by the action may have the "most good" or the best possible outcome.[44] Given that this study intends to construct an African population's ethics, it is appropriate to interpret "Person-Based Consequentialism"

[42] Falola, T., 2003. *The power of African cultures*. University Rochester Press.

[43] Melinda A. Roberts, "Person-Based Consequentialism and the Procreation Obligation," in The Repugnant Conclusion: Essays on Population Ethics, (Dordrecht: Springer, 2004), 100.

[44] Arrhenius, G., 2003. The person-affecting restriction, comparativism, and the moral status of potential people. *Ethical Perspectives*, *10*(3), pp. 185–195.

not strictly according to its utilitarian underpinning. But it is now to be decoded in terms of "Person-Based Theory" or "PBT." Person-Based Theory would mean that any conscious person, when taking actions, will have the duty to prefigure that there is necessarily a living or future person who will be directly affected by the actions at a given moment. In this way, the African ethics of intersubjectivity as the foundation of African population ethics would recommend acting only in a way that strengthens human interconnection for the populations of the moment and those of the future.

The main point that drives PBT within an African ethics of intersubjectivity is that "it must be seen as one aimed at a more fully human social order, A society in which all people are able to develop to their full potential."[45] This point must always accompany the African person's actions. Enlightened by PBT's concept, we are already confronted with the moral question of the actions by past peoples and the current populations with a quantitative and qualitative vision about the populations to come. Current populations' actions on the environment are characterized by various consequences, which unfold in the loss of forests, accelerated desertification in many parts of Africa, prolonged drought, scarcity of water, unusually torrential rains, natural disasters, and many other climatic ills.[46] We foresee that the people existing now and those to exist in future generations, whose common denominator would be the search for a maximum disposition for a better life, will not have equitable treatment insofar as the existing people have already made choices and performed actions that may disadvantage people of future generations. The issue of the welfare of future generations is crucial to population ethics in its ethical intersubjective perspective.[47]

Is it really this obvious that the existing populations are morally wanted based on their present and future actions vis-à-vis future populations? It appears obvious for the reason that, grounded on an ethical intersubjective perspective of population ethics, these conclusions are logically possible, though they may appear repugnant.[48] In order to prevent such obviousness, we must consider other theories, such as the telic sufficientarianism, which considers shortfalls from a sufficient level of welfare as morally bad as far as welfare is concerned.[49] Sketchily, sufficientarianism is referred to in political philosophy as an approach to distributive justice. It is "the doctrine advising the ethical observer to 'maximize the number of people who have

---

[45] Nhlanhla Mkhize, 2008: 33.

[46] Bujo, B., 2009. Ecology and ethical responsibility from an African perspective. In Murove, M.F (Ed.) *African Ethics. An Anthology of Comparative and Applied Ethics*. Pietermaritzburg, University of Kwazulu-Natal Press, 281–297.

[47] Blackorby, C., Bossert, W. and Donaldson, D., 2003. The axiomatic approach to population ethics. *Politics, Philosophy & Economics*, *2*(3), pp. 342–381.

[48] Huseby, R., Fourie, C. and Rid, A., 2017. Sufficiency, priority, and aggregation. *What Is Enough? Sufficiency, Justice, and Health*, pp. 69–84.

[49] Huseby, Robert. 2012. "Sufficiency and Population Ethics." *Ethical Perspectives* 19 (2): 187–206.

enough' in any situation."[50] In our view, beyond the aim of present populations to maximize their welfare, it should be projected that a certain level of well-being as an advantage should be aimed not only for a greater number of people but for all people, present and future.

It is not an egalitarian view of things that is being promoted here, but essentially a sufficiency principle which states that: "It is, in itself, bad if a person is not sufficiently well-off. It is worse the farther from a sufficient level a person is, and worse the more people that are not sufficiently well-off."[51] If all things are considered, the state of affairs of the person is such that he/she does not have well-being; such a state of affairs is bad in itself. When the person is far from the acceptable level of sufficiency in life, this is bad in itself. And the situation is even worse for anyone who is very far away in comparison to the situation of well-being that could be predicted. And it is even worse if there are many more people who do not have a sufficient level of well-being. Such a perspective aims to emphasize that in the present case, the attention is given to every individual because it would not be a question of maximizing the well-being of many people, even if, along the way, we would end up ignoring even a single individual. The principle of sufficiency takes into account each individual's well-being and everyone else's.

There are two forms of sufficientarianism: "Strong Sufficientarianism: Benefits that lift individuals above some threshold level T matter more than equally large benefits that don't (whether they occur above T or below it). Weak Sufficientarianism: Any benefit below T, no matter how small, and no matter how few individuals, outweighs any benefit above T, no matter how large, and no matter how many individuals. Below T equally broad benefits matter more the worse off the recipient is."[52] Taking these two together, strong sufficientarianism and weak sufficientarianism, what is remarkable is the fact that both approaches lead to the principle of telic sufficientarianism, which, in fact, has a sense of priority, such that the situation of people who are below the sufficiency level should be the first concern.[53] This is explained by the fact that, according to this principle, a world in which a single individual is insufficiently well-off is less good than a world in which everyone has a sufficient level of well-being. As evident as it appears, this approach is exposed to criticisms of the very principle of sufficientarianism, which I will not explore here. It is enough to note, for example, that if the technological development of existing populations increases their advantages in improving the quality of life and therefore increasing their well-being, such sufficiency outweighs any other development that does not lift the benefits of existing and future populations, leaving them below the sufficiency threshold. Contrastingly, a level of sufficiency that may be below a threshold T, but which still provides benefits for the existing populations, outweighs

---

[50] José Carlos R. Alcantud, Marco Mariotti, and Roberto Veneziani, "Sufficientarianism," Theoretical Economics 17, no. 4 (2022): 1529–57.

[51] Huseby, R., 2012. Sufficiency and population ethics. Ethical Perspectives, 19(2), pp. 180.

[52] Segall, S., 2016. What is the Point of Sufficiency? *Journal of Applied Philosophy, 33*(1), pp. 36–52. 37.

[53] Huseby, "Sufficiency and Population Ethics," 192.

any sufficiency above threshold T, regardless of the number of people concerned, because this seems to be the level at which welfare is possible for these concerned people.

From the above, it should be noted that the difficulty persists in passing a moral judgment. What counts the most: is it only to maximize the well-being of the populations in a situation such that the said populations may have a passable life (life worth living) instead of another life which would be better than the first one, an excellent life? It seems to us that more attention is paid to the concept of well-being without specifying the content of this concept in such a way that we find ourselves back at the starting point of the question itself. In our opinion, it is not enough to create more populations with just a passable life, but we need populations that have an acceptable and better life; a livable environment is therefore one of the characteristics of such an acceptable and recommendable life.

The previous elucubrations show sufficiently how much less easy it is to distinguish the level of morality of the actions of intergenerational populations from the point of view of benefits, sufficiency, and well-being because what appears to be well-being for an existing population may seem less of a well-being for another population to come. The point at hand, however, is that sufficientarianism opposes any action that, in aggregate, would leave populations significantly less well-off, even when large welfare gains are contemplated for other populations.[54]

In slightly clearer terms, and by means of illustration the Uganda and Tanzania Crude Oil Pipeline Project, a project presented as a lifesaving initiative for millions of Ugandan populations, considering the multiplier effects of the said project. For example, suppose Uganda initiates an oil project that requires the installation of pipelines to transport crude oil from the western region of Uganda, passing through Tanzania on the south side of Lake Victoria. Lake Victoria, passing through the Tanzanian towns of Sinyanga and Singida, finally arrives at the Tanzanian port of Tanga.[55] According to strong sufficientarianism, the state of affairs is that for the populations existing at these places, the benefits to be gained in relation to the situation of their well-being as a result of the economic effects of such a project are equal to the economic benefits of the project when these economic benefits would have maximized the number of individuals benefiting from improved well-being than before the project. This means that the project is morally credible if it has brought some relevant form of well-being to the existing populations. And, according to weak sufficientarianism, the benefits in relation to the well-being of the population even if the economic benefits do not maximize the well-being of individuals in relation to their pre-project situation, their level of well-being remains the same, because they were already at a threshold of enough well-being as a matter of the state of affairs; this means mean that the existing populations already possessed a life worth-living that has not been modified by the project. Our position is that

[54] Huseby, R., Fourie, C. and Rid, A., 2017.

[55] EACOP, Unlocking East Africa's Potential, 2023, accessed August 20, 2023, https://www.gem.wiki/East_African_Crude_Oil_Pipeline_(EACOP)

instead of considering only the telic sufficientarianism, which concerns only the final results with regard to the benefits in relation to well-being in the sense of maximizing the number of individuals whose well-being ends up improving, we suggest seeking telic sufficientarianism, which is more concerned with the means of evaluating the outcomes in terms of the state of affairs of persons. This is more significant when it comes to the actions of the populations present on the environment. It is not only a question of aiming at maximizing the well-being of existing populations by creating, for example, beneficial economic projects for a greater number of individuals; it is also necessary to consider the present duty of agents in relation to the good of future populations, actions on the environment must be imbued with a character of morality in order not to diminish the chances of well-being of future populations. The idea of telic value, which sufficientarianism must always assume, is vital, especially when it comes to human actions on the environment. This is because a level of sufficiency of the existing population that benefits millions can be morally bad if, in aggregate, it leaves future populations at a level of sufficiency that is less beneficial. In the ideal of considering the interventions of existing populations on the environment without critically creating environmental damage that impairs the well-being of future populations, one may have to reconsider the sacredness of the environment.

9.5 Reclaiming the African Sense of the Sacredness of the Environment for the Well-Being of Present and Future Populations

From the outset, I must draw attention to this sense of sacredness in relation to nature and the environment. This is not a return to animism or a religious cult that humans render to nature. Nor is it a question of deifying nature or the environment. It is an attitude of deep respect and regard for the very life of nature, which must be taken seriously as a unique way of life that needs to be respected and preserved for the very good of nature on which human life intrinsically depends. Traditional Africa was distinguished by this practical and applied intelligence in the face of nature and the environment.

In the relationship between modern African human beings and the environment, it is important to recognize the cultural and spiritual significance of the environment in African societies. Many African cultures have traditionally viewed the natural world as a source of spiritual connection and inspiration and have developed a deep respect for the land and its resources.[56]

[56] Falola, T., 2003. *The power of African cultures*. University Rochester Press.

In most parts of Africa, traditional populations had a natural and revered bond with nature, understanding themselves as part of the environment and relying on it for their livelihoods.[57]

In these modern days, the relationships and the sacred bond between humans, nature, and the environment appear no longer strong, except where some traditional and indigenous people strive to maintain some ties with nature for reasons of livelihoods or medicines, as is the case with rural populations of indigenous African peoples.[58] Such might also be the case for the San peoples of Southern Africa, who continue to maintain a revered connection and respect for nature through a belief that all living beings, intelligible and unintelligible, are interconnected.[59] Though still relying on hunting for food, they are known for their sustainable hunting practices, their system of gathering greens and collecting wood, as well as the construction of their habitat that minimizes the destructive impact on wildlife populations. Such traditional practices of sustainable hunting and food gathering, though not applicable on a large scale all over Africa and the world, prioritizing the well-being of wildlife and the environment over short-term economic gain, can serve as a mental model of environmental preservation to instill in existing African populations. As an outcome, these populations can develop an almost sacred connection with the environment, such that everyone may understand that the quality of human life depends exactly on the quality of human relationships with the environment.

Given the above thoughts, we remain aware of the decreasing belief in the sacredness of nature in modern Africa. Nonetheless, an African researcher is to be more of a possibility-thinker and less of a thinker of impossible possibilities in the sense that he/she must use the realities of the people and open their horizons to possibilities of solutions to the fundamental problems facing life in Africa. Appealing to the sense of sacredness that Africans had in nature is one of these horizons that still seems possible. This is because even if we note the decrease in attachment to certain values anchored in our African cultures, we must also recognize that there are still vestiges that, like old trunks of cut trees, can still bud some new plants. This amounts to saying that the sense of the sacred, which would mean an almost intuitive respect for nature and the environment, seems less present, but the African person has not completely departed from his ancestral and traditional values, which still have, to this day, a non-negligible existential significance. This concerns, for example, the importance accorded to initiation rites and rites of passage into adulthood in almost all the tribes across sub-Saharan and southern Africa.[60] In this context, we realize

---

[57] Leal Filho, W., Esilaba, A.O., Rao, K.P. and Sridhar, G., 2015. *Adapting African agriculture to climate change: Transforming rural livelihoods*. Springer. 1–23.

[58] Nathani-Wane, N., 2014. *Indigenous African knowledge production: Food-processing practices among Kenyan Rural Women*. University of Toronto Press.

[59] Andrew Brown Smith, The Bushmen of Southern Africa: A Foraging Society in Transition (Cape Town: New Africa Books, 2000).

[60] Lebese, R. T., Mothiba, T. M., Mulaudzi, M. T., Mashau, N. S., & Makhado, L., 2022. "Rite of passage: An African indigenous knowledge perspective." In: Mulaudzi, F.M. and Lebese, R.T.(Eds). *Working with indigenous knowledge: Strategies for health professionals*. AOSIS Publishing. Cape

that rites of passage, initiations into adult life, can constitute a basis for the revitalization of the beliefs in the sacredness of nature and the environment. Rites of passage do not take place in the middle of African towns or cities. They are most often held in mountains, scrublands, and forests, near rivers and streams. Interestingly, in the collective African imagination, these places are undoubtedly invested with a sacred character, all the more so since they are also, in factual or imagined ways, as a matter of African belief, the habitat of the spirits of life.[61] These places of initiation already symbolize the source of life, the reservoir of well-being, the temple of goodness. For the most part, these places command an almost intuitive respect. Hence, we must say that the young men and women who enter rites of passage are not a negligible handful of African populations. If initiation masters could inculcate to the newly initiated a focus on the respect that the environment must always benefit from humans, there is therefore reason to believe that these thousands of young African populations will come out of the initiations with another look at the environment and will act accordingly, because, among other things, "the Rites of passage create expectations for socially appropriate behaviors."[62] From these populations of new adults, we can jump-start the respect for the environment in the backdrop of the very sense of sacredness attached to the very places of rites of passage, toward a new impetus to re-educate the general population on the new attitude with regard to the environment as a reservoir of life.

It is a fact that humans and the environment are in a "give-and-take" relationship. If humans give to the environment everything that promotes and allows the well-being of the environment, in return, they receive from the environment that which will help to maximize their own well-being. If, on the contrary, they serve the environment with that which reduces its chances and possibilities of well-being, this is what humans will also receive from nature and the environment. This should be a light from the perspective of person-affecting moralities in its person-affecting restrictions approach.[63] Such thinking is eminently ethical insofar as we consider that well-being is good and it must be achieved for existing populations, and also that it must be projected for future populations. In reference to the environment, we can, in a certain way, affirm that having an attitude of reverence toward nature is good for the existing populations to the extent that such an environment is suitable for a healthy life, allowing the realization of their well-being. Consequently, we must also consider achieving for future populations this same type of well-being, which would result from an environment respected and well-treated by existing populations in an impersonal sense, because every individual of future populations

Town, South Africa, 51–68. Bujo, B., 2009. Ecology and ethical responsibility from an African perspective. In Murove, M.F (Ed.) *African Ethics. An Anthology of Comparative and Applied Ethics*. Pietermaritzburg, University of Kwazulu-Natal Press, 281–297, 282.

[61] Nwadiokwu, C.N., Nwadiokwu, E.S., Favour, E.N. and Okwuazun, M.E., 2016. Rites of passage African traditional region. *International Journal of Education and Research*, *4*(9), pp. 41–50. 45.

[62] Lebese, "Rite of passage' 55.

[63] Holtug, N., 2004. Person-affecting moralities. In *The repugnant conclusion: essays on population ethics* (pp. 129–161). Dordrecht: Springer Netherlands. 161.

would be included. It is therefore in this understandably sacred relationship with the environment that we envision such an ethics of well-being. The existing populations, by seeking what is good, their well-being, must at the same time create today the conditions of what will be good for the populations to come, ultimately, their well-being.[64] Nevertheless, in recent years, there has been an increase in urbanization and industrialization across Africa, leading to a growing disconnect between people and the natural world. This has led to a number of environmental challenges, including deforestation, soil erosion, and pollution, which can have significant impacts on both human health and the environment's well-being. From corner to corner of Africa, efforts are visible in line with preserving cultural heritage in connection with the environment for sustainable tourism around natural sites in Botswana,[65] which can also be extended to South Africa, Zimbabwe, and Zambia.

Despite challenges, there are numerous efforts underway to promote the sustainability of the environment in Africa, thus helping existing populations attain their well-being and projecting a welfare for future populations. For example, there is a growing number of organizations working to promote renewable energy, reduce waste, and improve conservation practices. Additionally, many African countries are working to integrate sustainable development practices into their economic planning, recognizing that protecting the environment is critical to ensuring long-term economic sustainability and the well-being of the majority of people without neglecting the crucial issue of overpopulation.

9.6 The African Birth Rate, Population, the Environment, and the Value of the Human Person

Reflecting on the cohesion between population ethics and the environment leads to anticipating the type of well-being that people can enjoy when there is indeed a real and active harmony between these two entities, of which one bears direct responsibilities towards the other. These responsibilities stem from the fact that of these two entities, one of which, the human being, has knowledge, intelligence, applicable laws, capacity for decisions, and moral obligation (reference is made to Aristotle's Ethics and the African intersubjective ethics explained above). While the other, the environment, has neither intelligence nor moral order, but whose laws are proper to its nature and must regularly be applicable by the order of nature itself (reference is made to St Thomas Aquinas's natural law tradition). It can be asked if the birth rate has any influence on the gain or negative results in the encounter of humans and nature? Thus, humans are to gain any benefits for their well-being when there is

[64] Holtug, N., 2004. Person-affecting moralities. In *The repugnant conclusion: essays on population ethics* (pp. 129–161). Dordrecht: Springer Netherlands.

[65] Brown, B.A., 2020. Integrating Cultural Heritage into Sustainable Tourism Development. *Botswana Cultural Heritage and Sustainable Tourism Development: A Handbook of Theory and Practice*. Cambridge Scholars Publishing. 182.

harmonious unity with the environment, or face disastrous impacts, once the true natural order of the environment's best existence is violated by the actions of humans.

Having obtained a clear and distinct picture of the true relationship between populations and the environment, it is now possible to reflect on aspects such as an event when the well-being of existing populations is seriously compromised by the lack of existential harmony between populations and the environment. Questions may be asked: do these populations have the right to invite into existence other populations whose well-being would also be compromised? What about future populations, who decides for them? These questions direct us toward a strict consideration of the birth rate in Africa and its impact on human' actions toward the environment. A quick scan of the African continent, according to the United Nations' Population Prospects 2022, offers the following: "the current birth rate for Africa in 2023 is 31,599 births per 1000 people, a 1.27% decline from 2022. The birth rate for Africa in 2022 was 32,005 births per 1000 people, a 1.25% decline from 2021. The birth rate for Africa in 2021 was 32,411 births per 1000 people, a 1.23% decline from 2020. The birth rate for Africa in 2020 was 32,816 births per 1000 people, a 1.22% decline from 2019."[66]

Even if the general trend shows a decline in the birth rate in the world, and in Africa in general, there is no doubt that the world population, which reached 8 billion this year, will exceed this figure in the next 50 years, and the African birth rate will not remain unchanged. It is generally admitted that the African birth rate is the fastest in the world.[67] Birth rate is one of the important factors that can contribute to population growth or population decline, economic growth or economic decline. The economic impact of a birth rate has a direct link with the environment because the growth of the population puts a strain on natural resources. High birth rate will mean excessive use of natural resources in Africa, while low birthrate can lead to a shortage of workforce as a consequence of the aging population, which can lead to unstable economic gains. This necessitates a balance between birth rate and environmental sustainability by all policy-makers so that a harmony is always maintained between the reduction of poverty among the existing populations and the protection of biodiversity.[68] The increase in the number of populations requires an increase in the creation of economic resources to ensure the well-being of humans, and this will necessarily have an impact on the environment. For a positive outcome, public policies must ensure that economic practices and factors do not create an advantage or disadvantage in favor of one or the other, but at all costs seek a vital

[66] Macrotrends, 2023. Africa Birth Rate 1950-2023. [Online]. Available from: https://www.macrotrends.net/countries/AFR/africa/birth-rate and. [Accessed 18 August 2023].

[67] Mungazi, Z.S., 1986. Population Size in Africa Countries; An Evaluation. Economics; Walters, S., 2021. African population history: Contributions of moral demography. *The Journal of African History*, *62*(2), pp. 183–200.

[68] Sandbrook, C. and Roe, D., 2010. Linking conservation and poverty alleviation: the case of great apes. *An overview of current policy and practice in Africa. The Poverty and Conservation Learning group/Arcus Foundation. 7.*

balance between the creation of wealth and the sustainability of the environment, all to the benefit of the welfare of present and future populations.

The growth in birth rate is important for the survival and flourishing of human civilization, and individual autonomy and well-being are also paramount. Reality shows that in countries with a high birth rate and a concentration of populations, there are few resources to share. Given the economic difficulties of most countries on the continent, this results in a very low level of well-being of populations who live below acceptable standards. This will be explicit in all areas of life, for example, sanitation in cities, pollution in cities and countryside, accelerated deforestation, lack of schools for an abundance of children, lack of jobs for many university graduates, overcrowded cities with few or non-existent social housing.[69] Under these conditions, well-being in terms of psychological satisfaction, economic gains, and good conditions of life obviously seems compromised by the precariousness that will characterize the lives of populations, and such a state of affairs will also impose strong demands on the environment, which will suffer more.

However, with the notion of life being very revered in Africa, sometimes efforts are made to distinguish less between the quantity and the quality of life of the populations. We would be satisfied with quantities, because life intrinsically is a fundamental value of the African person, so that what matters most is having a large family without worrying too much about the quality of life as long as each member of the family, clan, and village enjoys the vital act, such that any life is a life worth living.[70]

Reflecting on self-consciousness and interaction with the environment, should we accept the principle that, since life itself is a fundamental value, what matters most is the number of living populations, regardless of the level of their well-being, in the sense that there is nothing better than life itself? Derek Parfit contends that, "given the limits to the world's resources, we could not in fact produce the greatest possible sum of happiness, or the greatest amount of whatever makes life worth living, by producing an enormous population whose lives were barely worth living, merely technically impossible. In order to suppose it possible, we merely need to add some assumptions about the nature and availability of resources."[71] In my opinion, it is one of the advantages of self-consciousness to allow us to know ourselves better as human persons and to understand clearly that the best is not only to live, but to live better. This means that the quality of our life corresponds best with the dignity of the human person that we are, being clothed with reason, morality, and the ability to perfect our existence by constantly ascending toward the infinity of our possibilities. African population ethics will have its full meaning when it enlightens

[69] Parfit, D., 2004. Overpopulation and the Quality of Life. In *The repugnant conclusion: Essays on population ethics* (pp. 7–22). Dordrecht: Springer Netherlands; Bernard, F.E. and Thom, D.J., 1981. Population pressure and human carrying capacity in selected locations of Machakos and Kitui districts. *The Journal of Developing Areas*, *15*(3), pp. 381–406.

[70] Amadiume, I., 1997. *Re-inventing Africa: Matriarchy, religion and culture*. Zed Books; Davis, R.H., 2005. Encyclopedia of African history and culture. USA.

[71] Parfit, Overpopulation and the Quality of Life, 149.

African opinion on the need to establish a reasonable balance between the quality and quantity of life of Africans. In any way, the excess population has its significance and its importance, but when it comes to the quality of life, the excess should not be an element of diminishing the very dignity and value of the human person. Insofar as the excess, the essence of what would make the human being a worthy being and master of his existence, can be dissolved. Another essential element of the answer would be in the simple equation that the excess population is always a destructive element of the environment that could have permitted an existence worth bearing for existing populations, and in prospect for future peoples.

This can be justified if we analyze the population surplus from the axiomatic angle of population ethics. Particularly if we adopt the strong Pareto and the weak Pareto as axioms of our analysis.[72] In the strong Pareto, for the same population classified in categories x or y, for this same population, if the well-being of each individual is equally good, this would mean that the well-being of the two categories is good for these populations. Further, in the weak Pareto, if the well-being of each individual is not worse, this would mean that at least the well-being of a person in category x is good, and that, all things being equal, the welfare of populations in category x is good only in comparison to category.[73] In the applicability of this axiom, we will say that, in terms of the environment and the impact of climate change, for the same population currently existing in Africa, category x would be the countries that so far experience the least pangs of climate change, and category y would include all other countries. It is clear that not one country is spared from the horrors of climate change, which leads to the affirmation that, in the present case, even the axiom of strong Pareto does not hold true. In the case of weak Pareto, we can say that the well-being of the population of category y is certainly not better, but at least as bad as category x. If this were the case for all populations, this would mean that in both categories, the well-being of the populations does not reach the acceptable threshold of a life worth living. In clearer terms, for example, we have observed in the last decades that most African countries have experienced climatic disasters, of course, some more severe than others. Category x countries have climatic disasters at a level less than countries in category y. These countries may include countries like Zambia, Botswana, Lesotho, Cameroon, and many others; though experiencing climatic disasters not so severe, they still have in some ways climate change impacts that seriously impede the quality of life. Unlike category y countries, countries such as Mozambique, South Africa, Malawi, Madagascar, and many others with clearly deadly impacts of climatic changes, the quality of life is critically embedded.

Considering the welfare of the populations in the different categories of countries mentioned, how would we project the welfare of future populations? It is obvious that when we reflect on the number of present and future populations, we realize

---

[72] Blackorby, C., Bossert, W. and Donaldson, D., 2003. The axiomatic approach to population ethics. *Politics, Philosophy & Economics*, 2(3), pp. 342–381.

[73] Blackorby, C., Bossert, W. and Donaldson, D., 2003. The axiomatic approach to population ethics. *Politics, Philosophy & Economics*, 2(3), pp. 342–381. 364.

that if climatic disasters severely disturb the well-being of existing populations, this means that if the number of populations were to multiply, the result would be that many more people would suffer and have less well-being.[74] It further indicates that many more people would die without achieving a better standard of living for the reason, because if the existing populations are already in such climatic precariousness, if nothing is being done to change the state of affairs, the populations to be created will have more and more intense climatic precariousness and less and less of the well-being so sought after. Such a dark picture forces us to pull ourselves together now to finally consider what type of future populations we ought to create. We would agree that the best would be to create a population that is thrown into the world while open to self-consciousness and intersubjectivity with other humans and a dignified environment capable of accomplishing the population's well-being.

9.7 Conclusion

Cogitating on the attainment of the well-being of African populations, this study began with an applied hermeneutic of concepts such as self-consciousness, intersubjectivity, and environment. This is a perspective that seems realistic for a population ethics that suits Africans' longing for well-being. For a useful interconnectedness, each subject must have a deep awareness of the self and the other. This other refers both to the rational subject, which is the next-self among the human species, and to the non-rational subject, which is the environment. In light of this, consideration was rendered to the African ethical intersubjective perspective of population ethics. Meaning that in imagining the calling of future populations into existence, equally projecting the likelihoods of their well-being, we would reason that the anticipated situation is such that their well-being ought to be no less than that of existing populations, if not equal to or, better still, greater than that of current populations.[75] The Communitarian African ethics, here considered in the conceptual theme of the African ethics of intersubjectivity, underlining relationality and sociality, assumes that true self-consciousness necessarily leads to a type of conduct that honors the human essence of the other person. In this sense, an African population's ethics must open the community space to include the environment as a relational and social being, without which the community space for well-being and common good is seriously diminished. This calls to revere the maxim of African Populations Ethics. The Maxim stated in the previous section seems consistent with the African ethicists who have widely argued that, instead of individualizing well-being, Africans generally conceive it as a community-sought-after-good because no one

---

[74] Parfit, D., 2004. Overpopulation and the Quality of Life. In *The repugnant conclusion: Essays on population ethics* (pp. 7–22). Dordrecht: Springer Netherlands 147.

[75] Huseby, "Sufficiency and Population Ethics," 189.

can achieve well-being alone, or increase life force as an island.[76] We are not talking here about a kind of egalitarian well-being but about the principle of deontic sufficientarianism to think and act in a way to realize well-being not only for oneself, or the majority, but for all other people affected by one's actions. According to this principle, it is not to expect that each individual may have the same well-being, but the duty is to make sure that each person should be accorded the fair chances to reach their well-being.[77] This entails a regular frequency of sufficiency of the individuals' well-being so that every effort should be made to step up the total amount of well-being for the existing populations, as well as for other peoples to be brought into existence. To advance such an approach, we relied on the concepts of self-consciousness and African intersubjectivity. It seemed important to us that to imagine future populations whose welfare would be no less good than the well-being of the existing populations, we should align concomitantly the consciousness of self, the connectedness between subjects of the existing populations, and consciousness and interconnectivity with the environment. This is to foster the existential accolade as the basis of well-being for all. Self-consciousness, coupled with the maxim of the African ethics of intersubjectivity, brings about knowledge such that one understands henceforth that achieving intersubjectivity with the environment for existing populations and also with in mind the projection for future peoples is a basis of well-being.[78] In this approach, we exposed the accounts of the impacts of climate change in Africa. It is estimated globally that 12 million hectares of forests are destroyed annually, and this situation contributes up to 25% of greenhouse gas emissions.[79] Consequently, several regions of Africa have experienced very devastating natural disasters in recent years; among them, severe droughts, violent winds, destructive floods, forest fires, and the decimation of domestic animals due to drought and lack of vegetation for pasture. These scourges are part of the consequences of global warming around the world.

As far as Africa is concerned, with few financial means to contribute to the reduction of greenhouse gas emissions, each individual must be led, one way or another, immediately to limit practices that hasten deforestation due to agriculture and other harmful uses of the environment. Aware of the demands of the population's ethics

---

[76] Dube, Musa "I Am Because We Are,"2009; Mkhize, N. "Ubuntu and Harmony: An African Approach to Morality and Ethics"; 2008; Mafunisa, J. M. "Ethics, African Societal Values and the Workplace," 2008; Bujo, Bénézet. "Ecology and Ethical Responsibility from an African Perspective", 2009.

Dube, Musa "I Am Because We Are,"2009; Mkhize, N. "Ubuntu and Harmony: An African Approach to Morality and Ethics"; 2008; Mafunisa, J. M. "Ethics, African Societal Values and the Workplace," 2008; Bujo, Bénézet. "Ecology and Ethical Responsibility from an African Perspective," 2009.

[77] Blackorby, Charles, Walter Bossert, and David Donaldson. "The axiomatic approach to population ethics." *Politics, Philosophy & Economics* 2, no. 3 (2003).

[78] Huseby, Robert. "Sufficiency and population ethics." *Ethical Perspectives* 19, no. 2 (2012).

[79] United Nations Environment Program, "10 Ways You Can Help Fight the Climate Crisis," March 3, 2022, accessed July 17, 2023, https://www.unep.org/news-and-stories/story/10-ways-you-can-help-fight-climate-crisis

in its welfarist and axiomatic aspect, we should essentially think of the welfare of the peoples yet to be called into being. It is therefore imperative that each citizen of Africa, in the part of the Earth he/she inhabits, undertakes to reforest the surrounding area. May the churches and schools which have a predominant voice, together with environmentalists, political leaders, civil society, and activists, contribute to igniting in the minds and actions of current populations a sense of self-consciousness and the deontic appeal of the ethics of intersubjectivity connectedness with the environment. Thus, it bestows future populations with a sufficiency of opportunities for a better relationship with an environment conducive to well-being. I recommend that African populations be continually reminded of these words of Bujo Bénézet,

> The contemporary person living in a consumer-oriented society is compelled by an insatiable urge to produce more and faster by ever more sophisticated methods. This mode of production is not directed towards making the world human. Trapped by modern technology, the modern person strives to deal with nature as with a slave…The dramatic consequences of modern technology should encourage Africans not to abandon their cultural heritage too hastily…Climate change is here with more and more erratic weather patterns and negative effects on agriculture. Certain animals and plants are becoming victims of profit-making. Destruction of the rainforests is determined by Western "rationality" that does not value nature's mysteries and rejects its symbolic language…For human beings to live harmoniously depends on nature. Cosmic elements enable community life to function.[80]

References

Abraham, Willie E. 2015. *The mind of Africa*. African Books Collective.

Africa News. *South Africa: KwaZulu-Natal Floods*. Knowledge Hub. April 8, 2022. https://www.jbarisk.com/knowledge-hub/event-response/south-africa-kwazulu-natal-floods/. Accessed 4 November 2025.

African Center for Economic Transformation (ACET). n.d. *Human well-being in the African Transformation Index*. https://acetforafrica.org/ati/growth-with-depth/human-well-being/. Accessed 25 March 2024

Alcantud, José Carlos R., Marco Mariotti, and Roberto Veneziani. 2022. Sufficientarianism. *Theoretical Economics* 17 (4): 1529–1557.

Amadiume, Ifi. 1997. *Re-inventing Africa: Matriarchy, religion and culture*. Zed Books.

Aristotle. 1999. *Nicomachean ethics*. Translated by Terence Irwin. Indianapolis: Hackett.

Arrhenius, Gustaf. 2003. The person-affecting restriction, comparativism, and the moral status of potential people. *Ethical Perspectives* 10 (3): 185–195.

Bernard, Frank E., and Derrick J. Thom. 1981. Population pressure and human carrying capacity in selected locations of Machakos and Kitui districts. *The Journal of Developing Areas* 15 (3): 381–406.

Blackorby, Charles, Walter Bossert, and David Donaldson. 2003. The axiomatic approach to population ethics. *Politics, Philosophy & Economics* 2 (3): 342–381.

Blackorby, Charles, Walter Bossert, and David Donaldson. 2004. Critical-level population principles and the repugnant conclusion. In *The repugnant conclusion: Essays on population ethics*, 45–59. Dordrecht: Springer Netherlands.

[80] Bujo, Bénézet. "Ecology and Ethical Responsibility from an African Perspective," 2009: 288–291.

Branco, Angela Uchoa, and Maria Cláudia Lopes-de-Oliveira, eds. 2017. *Alterity, values, and socialization: Human development within educational contexts*. Vol. 6. Springer.

Brown, Byron A. 2020. Integrating cultural heritage. In *Botswana cultural heritage and sustainable tourism development: A handbook of theory and practice*, 182. Cambridge Scholars.

Buber, Martin. 1970. *I and thou*. Vol. 243. Free Press.

Bujo, Bénézet. 2009. Ecology and ethical responsibility from an African perspective. In *African ethics: An anthology of comparative and applied ethics*, ed. Munyaradzi Felix Murove, 281–297. Pietermaritzburg: University of KwaZulu-Natal Press.

Davis, R. Hunt. 2005. *Encyclopaedia of African history and culture*. Facts on File.

Decety, Jean, and Jessica A. Sommerville. 2003. Shared representations between self and other: A social cognitive neuroscience view. *Trends in Cognitive Sciences* 7 (12): 527–533.

Dietz, Ton, and Els Veldhuizen. 2004. Population dynamics: An important intervening variable. In *The impact of climate change on drylands: With a focus on West Africa*, 79–82. Dordrecht: Springer Netherlands.

Dietz, Ton, David Millar, Saa Dittoh, Francis Obeng, and Edward Ofori-Sarpong. 2004. Climate and livelihood change in North East Ghana. In *The impact of climate change on drylands: With a focus on West Africa*, 149–172. Dordrecht: Springer Netherlands.

Dube, Musa W. 2009. 'I am because we are': Giving primacy to African indigenous values in HIV and AIDS prevention. In *African ethics: An anthology for comparative and applied ethics*. Scottsville, South Africa: University of KwaZulu-Natal Press.

East African Crude Oil Pipeline (EACOP). n.d. *Unlocking East Africa's potential*. https://www.gem.wiki/East_African_Crude_Oil_Pipeline_(EACOP). Accessed 20 August 2023.

Falola, Toyin. 2003. *The power of African cultures*. University of Rochester Press.

Fourie, Carina, and Annette Rid, eds. 2016. *What is enough? Sufficiency, justice, and health*. Oxford University Press.

Gillett, Grant R., and John McMillan. 2001. *Consciousness and intentionality*, 1–275. John Benjamin Publishing.

Gottlieb, Paula. 2015. Aristotle: Nicomachean ethics. In *Central works of philosophy v1*, 46–68. Routledge.

Hazell, Peter. 2004. Climate change and management of catastrophic risk. In *The impact of climate change on drylands: With a focus on West Africa*, 385–395. Dordrecht: Springer Netherlands.

Heidegger, Martin. 2002. *On time and being*. University of Chicago Press.

Holtug, Nils. 2004. Person-affecting moralities. In *The repugnant conclusion: Essays on population ethics*, 129–161. Dordrecht: Springer Netherlands.

Hoy, David Couzens. 2012. *The time of our lives: A critical history of temporality*. MIT Press.

Huber, Bernhard A., Bradley J. Sinclair, and Karl-Heinz Lampe. 2005. *African biodiversity: Molecules, organisms, ecosystems*. Springer.

Huseby, Robert. 2012. Sufficiency and population ethics. *Ethical Perspectives* 19 (2): 187–206.

Husserl, Edmund. 2014. *Ideas for a pure phenomenology and phenomenological philosophy: First book: General introduction to pure phenomenology*. Hackett Publishing.

Keromnes, Gaëlle, Sylvie Chokron, Macarena-Paz Celume, Alain Berthoz, Michel Botbol, Roberto Canitano, Foucaud Du Boisgueheneuc, et al. 2019. Exploring self-consciousness from self-and other-image recognition in the mirror: Concepts and evaluation. *Frontiers in Psychology* 10:719.

Leal Filho, Walter, Anthony O. Esilaba, Karuturi P. C. Rao, and Gummadi Sridhar. 2015. *Adapting African agriculture to climate change: Transforming rural livelihoods*. Springer.

Lebese, R. T., T. M. Mothiba, M. T. Mulaudzi, N. S. Mashau, and L. Makhado. 2022. Rite of passage: An African indigenous knowledge perspective. In *Working with indigenous knowledge: Strategies for health professionals*, ed. F. M. Mulaudzi and R. T. Lebese, 51–68. Cape Town, South Africa: AOSIS Publishing.

Macrotrends. n.d. *African birth rate 1950–2023*. https://www.macrotrends.net/countries/AFR/africa/birth-rate. Accessed 18 August 2023.

Mafunisa, J. M. 2008. Ethics, African societal values and the workplace. In *Persons in community: African ethics in a global culture*, ed. Ronald Nicolson, 111–124. Pietermaritzburg: University of KwaZulu-Natal Press.

Meissner, Darius. n.d. *Population ethics and theories of well-being*. https://www.utilitarianism.net/population-ethics. Accessed 2 September 2023.

Méjean, Aurélie, Antonin Pottier, Marc Fleurbaey, and Stéphane Zuber. 2020. Catastrophic climate change, population ethics and intergenerational equity. *Climatic Change* 163 (2): 873–890.

Merleau-Ponty, Maurice. 1964. Relationships with others in children. *Psychology Bulletin* 18:295–336.

Mkhize, N. 2008. Ubuntu and harmony: An African approach to morality and ethics. In *Persons in community: African ethics in a global culture*, ed. Ronald Nicolson, 35–44. Pietermaritzburg: University of KwaZulu-Natal Press.

Mungazi, Zibeon S. 1986. *Population size in Africa Countries: An evaluation*.

Munyaka, M., and M. Motlhabi. 2009. Ubuntu and its socio-moral significance. In *African ethics: An anthology of comparative and applied ethics*, ed. M. F. Murove, 63–84. Pietermaritzburg: University of KwaZulu-Natal Press.

Murove, M. F. 1999. *The Shona Concept of Ukama and the process philosophical concept of relatedness, with special reference to the ethical implications of contemporary neo-liberal economic practices*. PhD diss., University of Natal, Pietermaritzburg.

Nelson, Robert H. 2003. Environmental colonialism: "Saving" Africa from Africans. *The Independent Review* 8 (1): 65–86.

Nicolson, Ronald, ed. 2008. *Persons in community: African ethics in a global culture*. Pietermaritzburg: University of KwaZulu-Natal Press.

Nwadiokwu, C. N., E. S. Nwadiokwu, E. N. Favour, and M. E. Okwuazun. 2016. Rites of passage African traditional region. *International Journal of Education and Research* 4 (9): 41–50.

Parfit, Derek. 2004. Overpopulation and the quality of life. In *The repugnant conclusion: Essays on population ethics*, 7–22. Dordrecht: Springer Netherlands.

Piaget, Jean. 1936. *The birth of intelligence in children*. Neuchâtel: Delachaux and Niestlé.

Pons, F., and P. Harris. 2001. Piaget's conception of the development of consciousness. *Human Development* 44 (4): 220–227.

Richardson, Nigel. 2009. Can Christian ethics find its way, and itself, in Africa? In *African ethics: An anthology of comparative and applied ethics*, ed. M. F. Murove, 129–154. Pietermaritzburg: University of KwaZulu-Natal Press.

Roberts, Melinda A. 2004. Person-based consequentialism and the procreation obligation. In *The repugnant conclusion: essays on population ethics*, 99–128. Dordrecht: Springer Netherlands.

Ryberg, Jesper, and Torbjörn Tännsjö. 2004. *The repugnant conclusion: Essays on population ethics*. Dordrecht: Springer Netherlands.

Sandbrook, Chris, and Dilys Roe. 2010. *Linking conservation and poverty alleviation: The case of great apes. An overview of current policy and practice in Africa. The poverty and conservation learning group/Arcus Foundation*. 94pp.

Segall, Shlomi. 2016. What is the point of sufficiency? *Journal of Applied Philosophy* 33 (1): 36–52.

Shutte, Augustine. 2008. African ethics in a globalizing world. In *Persons in community: African ethics in a global culture*, ed. Ronald Nicolson, 15–34. Pietermaritzburg: University of KwaZulu-Natal Press.

Smith, Andrew Brown. 2000. *The Bushmen of Southern Africa: A foraging society in transition*. New Africa Books.

United Nations Environment Program. *10 ways you can help fight the climate crisis*. March 3, 2022. Accessed July 17, 2023. https://www.unep.org/news-and-stories/story/10-ways-you-can-help-fight-climate-crisis

Wallace, Benjamin, and Robert G. Kunzendorf. 2000. *Individual differences in conscious experience*, 1–424. Amsterdam: John Benjamins.

Walters, Sarah. 2021. African population history: Contributions of moral demography. *The Journal of African History* 62 (2): 183–200.

Wiredu, Kwasi, and Kwame Gyekye. 1992. *Person and community: Ghanaian philosophical studies I*. Vol. 1. Charlottesville, VA: The Center for Research in Values and Philosophy.
Zahavi, Dan. 2014. *Self and other: Exploring subjectivity, empathy, and shame*. Oxford: Oxford University Press.

Chapter 10
Population Boom Versus Scarcity of Resources in Malawi: An African Historical-Ethical Critique

Pascal Newbourne Mwale and Cobbener W. Sungani

Abstract A rapid increase in the population of a nation, or 'population boom,' poses a serious challenge to policymakers in both government and non-governmental organizational sectors, most especially in a country where there are few resources, for example, food, to cater for the rapid increase. The population outstrips the resources available. Historically, this book chapter will first track the changes in the population size of Malawi from colonial times to the present in order to demonstrate that the country continues to experience a population boom. Ethically, this book chapter will further demonstrate that the mechanisms for the distribution of scarce resources have not worked to the advantage of everyone who is most in need within this booming population. There has been no fair distribution of scarce resources in this nation. As the population grows, the inequality gap between the rich and the poor has also been increasing alarmingly, leading to the sheer neglect of the poor who end up suffering from many evils, including, but not limited to, chronic hunger, disease, poverty, poor education, no access to quality water, poor housing, no access to green energies, little to no market systems for cash, staples and other crops, and poor transportation systems. Drawing on Ubuntu philosophy, we will argue that these scarce resources can be made available and accessible to everyone who is categorized as poor. Ubuntu philosophy's Afro-communitarian and public morality, because of its 'other-regarding' morality of duties and engenders communal relationality and friendliness, can mitigate the aforementioned evils that are rooted in individualism, ensuring that everyone leads a good life.

Keywords Population boom · Scarce resources · Ubuntu philosophy · Malawi

---

P. N. Mwale (✉) · C. W. Sungani
University of Malawi, Zomba, Malawi
e-mail: pmwale@unima.ac.mw

B. Okyere-Manu, P. Mwipikeni (eds.), *An African Approach to Population Ethics*, Advancing Global Bioethics 23,
https://doi.org/10.1007/978-3-032-24464-2_10

10.1 Introduction

The increase in population—the so-called "population boom"—has been an issue of great concern across the globe. This chapter argues that a rapid increase in a nation's population poses a serious challenge to policymakers, in both government and non-governmental organizational sectors, especially in a country where there are few resources, such as food, to cater to the country's populace. In Malawi, the population greatly outpaces the resources available. For example, since the drought-accelerated famine of 2001, Malawi has not been able to produce enough maize, the staple crop, to satisfy the nation's appetites. As a result, the levels of hunger have remained serious year on year. Recently, in 2023, the Global Hunger Index (GHI) ranked Malawi at 88th of the 125 countries where sufficient data was collected globally; in this GHI survey, Malawi scored 21.1, meaning that in that year Malawi's level of hunger was ranked as 'serious.'[1] The state welfarist mechanisms for the distribution of scarce resources have not worked to the advantage of all of those most in need in this booming population. There has not been a fair distribution of scarce resources in this nation. As the population grows, the inequality gap between the rich and the poor has also been increasing alarmingly, leading to the sheer neglect of the ultra-poor who end up suffering from many evils including, but not limited to, chronic hunger, disease, poverty, poor education, no access to quality potable water, poor housing, no access to green energies, little to no market systems for cash, staples and other crops, and poor transportation systems. Using the Ubuntu philosophy, the chapter argues that these scarce resources can be made available and accessible to everyone who is categorized the least advantaged, or, simply, on this side of the world, the poor, if the rich exhibit the 'other-regarding' morality of duties or the public morality of communal relationality and friendliness.

10.2 Nyasaland's Population Projections, 1901–1977: A Stocktaking

Malawi, then Nyasaland, became a British colony in 1891. Prior to its colonization, the country had a very small population despite registering a number of immigrants from other parts of Africa. Various tribes encroached on the land of the original inhabitants, the Akafula or ABatwa, for a variety of reasons, ranging from warfare to droughts, which led to famine. During the 13th and the 16th centuries AD, for instance, the Maravi people, who had migrated from Zaire (now the DRC) through Tanganyika (now Tanzania), settled in the Central and Southern regions of what is now known as Malawi. Those who settled in the Central region and Southern region

[1] See https://www.globalhungerindex.org

are now known, respectively, as Chewa and Mang'anja/Nyanja.[2] The other groups of immigrants were the Tumbuka, Tonga, Ngonde, and Lambya, who eventually settled in the Northern region of Malawi.[3] Another group, the Ngoni, who had migrated from South Africa and Swaziland (now Eswatini), settled in various parts of Malawi around 1855.[4] It was in this same 19th century that the Yao, Lomwe, Sena, and, indeed, Europeans also arrived in what much later became Malawi.[5] It is not the purpose of this chapter to discuss such migratory flows and the reasons for various groupings' migration into the country. That lies beyond its scope. It is significant to note, however, that these cases of immigration significantly increased the initial population of the later British colony known then as Nyasaland (in 1891). Following the British colonization of Nyasaland in 1891, the country witnessed another increase in population due to the arrival of missionaries, traders, and other Europeans. In addition to the early immigrants discussed above, such as the Chewa, Mang'anja/Nyanja, Tumbuka, Tonga, Ngonde, Lambya, Yao, Ngoni, and Alomwe, among others, who had merged to live alongside the original inhabitants, the missionaries and other Europeans who came further significantly increased the population. In order to trace trends in the population of the land, the authorities there began to compile a Census.

The history of census taking in Malawi dates back to as far as 1891. The African population was included in the census for the first time in 1901. Since then, additional censuses have been conducted in a number of years: in 1911, 1921, 1926, 1931, 1945, 1956, 1966, and 1977.[6] Between the first census in 1901 and 1945, the population of Malawi increased by an average of 2.5% per annum. Evidence suggests that the population of Malawi experienced a faster rate of growth after 1945. The indication for the inferences comes from the growth rates of 2.7% and 2.9% for the periods 1911 to 1966 and 1931 to 1966, respectively.[7]

---

[2] Kalinga, O and Pike, J. "A Pre-Colonial History of Malawi." *The Nyasaland Journal 18, no.* 1(2000): 22–54.

[3] Kalinga, O and Pike, J. "A Pre-Colonial History of Malawi." *The Nyasaland Journal 18, no.* 1(2000): 22–54. See also Kalinga, Owen. "Trade, the Kyungus, and the Emergence of the Ngonde Kingdom of Malawi." *The International Journal of African Historical Studies* 12, no. 1(1979): 17–39.

[4] Thompson, T. "The Origins, Migration and Settlement of the Northern Ngoni." *The Society of Malawi Journal 34, no.* 1 (1981): 6–35.

[5] Thompson, T. "The Origins, Migration and Settlement of the Northern Ngoni." *The Society of Malawi Journal 34, no.* 1 (1981): 6–35.

[6] Malawi Population Census. Analytical report Volume 2 Infertility, Mortality and Population Projections 1901-2000, 1977.

[7] Malawi Population Census. Analytical report Volume 2 Infertility, Mortality and Population Projections 1901-2000, 1977.

10.3 Malawi's Population Since 1945: Provisional Totals

The full census of Malawi taken in 1945 recorded a total population of 2,049,914. This total, projected at an annual increment of 2.2%, would have produced an estimate of 2,983,600 in the year 1962. However, by 1955, the population of Malawi had already hit 3,240,318.[8] The trend continued in later decades: by 1975, the population had hit 5,280,965 with a growth rate of 2.86%; and in 1995, that is, after another two decades, the population had almost doubled again, being recorded at 10,112,816. Although in other years the population growth had indicated a small growth and sometimes a negative growth rate, for instance, the population recorded in 1993 was 10,257,233 and had a growth rate of 1.40%, the following year, 1994, had a – 1.22% growth rate with a population of 10,131,799. Similar trends were also recorded in the years from 1995 to 1997, with –0.19%, 1.96%, and then 1.96% again, respectively.[9] This trend was only apparent in a number of those years, mainly as a result of the scourge of HIV/AIDS which killed many people during the period. However, the population of Malawi has been growing rapidly despite various international and national campaigns aimed at controlling the population, through various family planning methods. At the time of writing, it has exceeded 20 million people (see the table for the trends of population growth in Malawi since 2010).

From 2010 to the present, the population of Malawi has increased insidiously. Of the 14,718,422 people that were recorded in 2010, despite registering several deaths following the COVID-19 pandemic of 2020 and 2021 that claimed many lives, and other diseases that have witnessed a surge in the death toll, the birth rate continues to outstrip the death rate. As specified in Table 10.1, the projections exemplify the population boom in Malawi since 2010. According to the World Bank report from 2020, Malawi's population has been growing at an average rate of 2.75% per annum, from 3.7 million people in 1960 to 18.1 million in 2018.[10] This growth rate, we argue, is very rapid and hence very difficult to sustain and it has accelerated the cost of living in Malawi to the detriment of the poor, who are already struggling with the problems of the widening inequality gap prevalent in the country. The appalling condition of the poor has been aggravated by politics, with bad governance characteristic of all regimes since independence (in 1964). Rampant public corruption, which has meant the wanton diversion of public funds into the pockets of greedy individuals, has become entrenched and remains out of control. The percentage of the poor in the country is the largest of all Malawi's population segments, and it has led to many struggling to meet basic needs. The distribution of resources by the

---

[8] World Bank. United Nations Population Division, World Population Prospects: 2022 revision, 2020.

[9] World Bank. United Nations Population Division, World Population Prospects: 2022 revision, 2020.

[10] World Bank. United Nations Population Division, World Population Prospects: 2022 revision, 2020.

Table 10.1 Malawi's population growth from 2010 to 2023

Year	Population growth rate	Growth rate (%)
2023	20,931,751	2.58
2022	20,405,317	2.59
2021	19,889,742	2.65
2020	19,377,061	2.70
2019	18,867,337	2.72
2018	18,367,883	2.72
2017	17,881,167	2.73
2016	17,405,624	2.76
2015	16,938,942	2.80
2014	16,477,966	2.83
2013	16,024,775	2.85
2012	15,581,251	2.87
2011	15,146,094	2.91
2010	14,718,422	2.93

The population of Malawi has increased since 2010
(Extracted from the United Nations World Population Division. World Population Prospects, 2022 revision)

government has failed to address the inequalities, hence relegating the poor further to abject poverty, to the detriment and peril of the rural ultra-poor. The next section of this chapter analyses the inequality gap between the rich and the poor as exacerbated by the badly managed and grossly unequal distribution of scarce resources from government interventions.

10.4 The Inequality Gap vis-à-vis the Poverty Rate in Malawi

In common with other Sub-Saharan African countries, Cornia and Martorano (2015) observe that tracing the long-term evolution of inequality in Malawi is problematic because no data are available for the period 1985–1993.[11] Equally problematic is that in the case of other periods, the data originate from different sources that use different income concepts and focus on different segments of the income distribution. It is, however, safe to point out that, unlike in the colonial and post-colonial periods, pre-colonial Malawi had low-income inequality because the country was subscribing to the egalitarian distribution of resources such as farmland, in accordance with its communal land tenure system. Predominant during this precolonial period was subsistence farming under general conditions of abundant supply.

[11] Cornia, G., and Martorano, B. "Building the IID-SSA inequality dataset and the seven sins of inequality measurement in SSA." Mimeo, UNDP Project on SSA Inequality, 2015.

Additionally, mining and industry were minimally developed in Nyasaland before the advent of colonialism.[12]

During the colonial and post-colonial eras, there was and has been a massive widening of Malawi's income inequality gap between the rich and the poor. Evidence suggests that inequality increased in the aftermath of colonization following the European annexation of African lands, which were previously held under a communal or customary system of land ownership. These European settlers, Harvey J. Sindima argues, introduced a plantation economy system to maximize the profits from exports. This period witnessed a growth in the cultivation of certain key crops: tobacco, cotton, coffee, and tea.[13] There are a wide range of factors that have contributed to this widening income inequality during the colonial period. Shawo Mwakilama has perceptively observed that the structure of the economy or productive system during the colonial period was geared toward extracting as much as possible from the colony. This deployment of resources and labour, so she argues, benefited the white settlers at the expense of the locals (the African natives), thereby worsening conditions for the country's native population: "The land dispossessions pushed Africans to the marginal and infertile pieces of land, something that resulted in food insecurity among the locals, who during the pre-colonial era were able to produce in abundance."[14] In addition, the colonial administration created a labour reserve as a source of cheap manpower for the mines of Rhodesia and South Africa. This deprived the communities of able-bodied men, leading to a significant decline in the country's productive capacity. What's more, the taxation policies that were introduced by the colonial administration of Nyasaland exacerbated the living conditions of Africans, thereby further widening the inequality gap.[15] In the post-colonial era, despite the high hopes that income inequality would be lowered following the nationalistic zeal that African leaders had demonstrated during the period of transition to independence, the inequality gap has further widened and spiraled out of control. Among this first generation of nationalistically zealous African leaders were Kwame Nkrumah of Ghana, Nnamdi Azikiwe of Nigeria, Ahmed Sekou Toure of Guinea, Leopold Sedar Senghor of Senegal, Julius Nyerere of Tanzania, Jomo Kenyatta of Kenya, Kenneth Kaunda of Zambia, Seretse Khama of Botswana, and Hastings Kamuzu Banda of Malawi.

Between 1985 and 1991, for instance, the inequality gap widened due to the decline of maize per capita of the period, and indeed, *inter alia*, the decline in net barter terms of trade recorded by the country until 1994, and to the distributional

---

[12] Cornia, G., and Martorano, B. "Building the IID-SSA inequality dataset and the seven sins of inequality measurement in SSA." Mimeo, UNDP Project on SSA Inequality, 2015.

[13] Sindima, H. *Malawi's First republic: An Economic and Political Analysis*. University Press of America, 2002.

[14] Shawo Mwakilama, Poverty and Inequality and the Quest for Inclusivity: A case study of Malawi at 54 Years of Independence, *American Journal of Social Sciences 6*, no. 4 (2018): 65–72.

[15] Shawo Mwakilama, Poverty and Inequality and the Quest for Inclusivity: A case study of Malawi at 54 Years of Independence, *American Journal of Social Sciences 6*, no. 4 (2018): 65–72.

impact of the global financing institutions' Structural Adjustment Programmes.[16] Although inequality reduced between 1993 and 2005 as a result of factors such as the introduction of the Starter Pack Programme for peasant and small-scale farmers, a decline in HIV/AIDS prevalence, and the stabilization of international terms of trade, the inequality gap between the rich and the poor widened dramatically. Since 2005, levels of income inequality have worsened due to the shift from low-inequality agriculture to high-inequality sectors such as services, commerce, transport, and construction. It was also driven by urban migration.[17] Cornia and Martorano further observed that between 2005 and 2011, the Gini coefficient, which measures the relative level of inequality in a nation, rose by 6 or 7 points.[18] Similarly, in their study of income inequality in Malawi, Mussa and Masanjala have argued that there is an increasing polarization in consumption at the extreme ends of the wealth distribution in Malawi. A comparison of the richest and the poorest 10% of the country's population paints an even more gloomy picture of growing inequality.[19] The consumption of the richest 10% was around 22 times higher than that of the bottom 10% in 2004; by 2011, however, it stood at 34 times higher, a clear indication that the Gini coefficient had worsened over this period.[20]

Notably, most people's income levels are way below the poverty line.[21] The Poverty Report of 2020 shows that in 2019/20, the poorer parts of the population had a consumption of 17.0% below the poverty line. Evidence from this report shows that poverty was deeper in rural areas (at 19.3%) than in urban areas (4.4%).[22] Similarly, there was a gap between the levels of poverty in urban areas, where it stood at 1.5%, and that observed in rural areas (8.7%). In terms of inequality at the national level, the report[23] indicates 0.423 in 2016/17 and 0.379 in 2019/20. Although the findings in the report indicate a decline of inequality from 0.423 to 0.379, this decline is negligible as far as inequality is concerned.

To remedy the challenge of the income inequality gap in Malawi, the government introduced programmes such as the above-mentioned Starter Pack Programme and the Farm Input Subsidy Programme (FISP). Disappointingly, however, these two anti-hunger/-poverty programmes had no positive impact. The inequality gap continued to widen unabated, and poverty levels worsened in subsequent years. Social cash transfers and other anti-poverty strategies have also had no positive impact. Mangani and others observe correctly that agricultural input subsidies have

[16] Pryor, F. Income Distribution and Economic Development in Malawi. WDP 36. World Bank, 1988.

[17] Cornia, G., and Martorano, B. "Building the IID-SSA inequality dataset and the seven sins of inequality measurement in SSA." Mimeo, UNDP Project on SSA Inequality, 2015.

[18] Cornia, G., and Martorano, B. "Building the IID-SSA inequality dataset and the seven sins of inequality measurement in SSA." Mimeo, UNDP Project on SSA Inequality, 2015.

[19] Mussa, R. and Masanjala, W. A dangerous divide: The state of inequality in Malawi. Oxfam, 2015.

[20] Mussa, R. and Masanjala, W. A dangerous divide: The state of inequality in Malawi. Oxfam, 2015.

[21] National Statistical Office. Malawi Poverty Report of 2020. Zomba, 2021.

[22] National Statistical Office. Malawi Poverty Report of 2020. Zomba, 2021.

[23] National Statistical Office. Malawi Poverty Report of 2020. Zomba, 2021.

dominated the agriculture budget at the expense of growth fundamentals such as agricultural research and development (agri-R & D), which could have otherwise significantly narrowed the prevailing inequality gap. Instead, agricultural input subsidies and other anti-poverty strategies and interventions take away a large chunk of the national annual budget and yet they make little impact, dismally failing to close the prevalent inequality gap.[24]

In addition to income inequalities, other inequalities are prevalent in the country: these include health inequalities, which are relatively high in Malawi. High-quality private clinics, for example, are a preserve of the rich and affluent families, while the poor are structurally barred access to these high-quality private clinics on the grounds of cost.[25] The primary healthcare facilities in Malawi, which are attended by most poor people, are known for their persistent acute shortages of medicines and staff, and hence expose the poor to extremely poor-quality healthcare services. For example, Mussa and Masanjala have argued that Christian Health Association of Malawi (CHAM) hospitals provide roughly 40% of the available healthcare in Malawi. However, their facilities vary hugely in terms of quality and are often allocated in very rural areas, but they do charge user fees, to the disadvantage of the rural poor who cannot afford such expensive private healthcare and medications. As a result of these charges, the poorest are excluded from accessing CHAM healthcare and often must travel much longer distances in order to access free public healthcare facilities.[26]

There are also gross inequalities in terms of access and, indeed, varying quality in the area of education. At all levels, Mussa and Masanjala observe, education is unequally distributed to the disadvantage of the poor. In Malawi, they argue, the level of inequality in education is more pronounced in the tertiary education sector, where a significant number of children from poor backgrounds fail to secure spaces in both private and public universities. Most of the high-quality secondary schools in Malawi, institutions which have qualified teachers and adequate levels of education materials, are dominated by children from affluent backgrounds. By contrast, most children from poor backgrounds end up in Community Day Secondary Schools (CDSSs) where resources and materials are severely in short supply. CDSSs are staffed by under-qualified and unqualified teachers. This has an impact on performance in the Malawi School Certificate of Education (MSCE) national examinations, where those in good-quality schools and from advantaged backgrounds score high grades and easily secure spaces in both public and private universities. In reality, private universities are unavailable to children from poor backgrounds. Hence, educational qualifications are unequally distributed in favour of the rich and the

[24] Mangani, R, Jayne, T., Hazel, Muyanga, P.M., Chimatiro, S., Burke, W. & Johnson, M. Agricultural Transformation in Malawi: Call to Action. *MWAPATA Institute Working Paper* No. 20/01, 2020. See also Cornia, G., and Martorano, B. "Building the IID-SSA inequality dataset and the seven sins of inequality measurement in SSA." Mimeo, UNDP Project on SSA Inequality, 2015.

[25] Mussa, R. and Masanjala, W. A dangerous divide: The state of inequality in Malawi. Oxfam, 2015.

[26] Mussa, R. and Masanjala, W. A dangerous divide: The state of inequality in Malawi. Oxfam, 2015.

affluent.[27] Mbewe and Kholowa argue that the Malawi education system provides unequal educational opportunities to the children of the urban rich and the rural poor through the system of Conventional Secondary Schools (CSSs) and Community Day Secondary Schools (CDSSs), with CDSSs offering low-quality education to the rural poor and CSSs offering high-quality education to the urban rich. Consequently, they argue, the children of the poor do not perform as well as children of the rich in national examinations. This has a knock-on effect in terms of their later career prospects, employment statuses, and general performance later in everyday life as adults.[28]

Moving on to another sector, we can outline how, in the area of agriculture, for example, from the early 1960s onwards, the state of Malawi switched to an estate and medium-scale farmers' strategy, thereby abandoning its initial policies which had involved public support for smallholder farmers.[29] This change in agricultural policies saw the state reallocating land leases, production licenses, subsidized credit, and fertilizers to estates and also to medium-sized farms, hoping to raise yields and introduce new farmwork methods. While this strategy registered success for 10–15 years, Pryor argues, the impact was grave, since it reinforced inequalities within the agriculture sector. This was evidenced by the squeezing of land, labour, and profits out of the smallholder farming sector, and also the decline in rural wages derived from smallholder farming.[30] Since then, the inequality in the agriculture sector has been widening, increasing discrimination against rural poor smallholder farmers.

In spite of several efforts to eradicate extreme poverty in Malawi by the government and, indeed, with the occasional and oft-intermittent budgetary support of Western donors and other relevant development partners, levels of poverty in Malawi are high and continue to rise.[31] Of course, hitherto the government has made various attempts to curb poverty in the country. However, the government's strategies in relation to economic growth, infrastructure development, and the provision of basic social services have so far proven to be unsuccessful. These strategies, as argued by Mussa and Pauw, include: the Poverty Alleviation Programme of 1994; the Malawi Poverty Reduction Strategy of 2002 to 2005; and, more recently, the Malawi Growth and Development Strategy (MGDS) (2006–2011 and 2011–2016).

---

[27] Mussa, R. and Masanjala, W. A dangerous divide: The state of inequality in Malawi. Oxfam, 2015.

[28] Mbewe, S. M. and Kholowa, F. "The Malawi education system and Reproduction of Social Inequalities: implications of Education for sustainable Development." *Malawi Journal of Education and Development 3,* (2010): 63–74.

[29] Pryor, F. Income Distribution and Economic Development in Malawi. WDP 36. World Bank, 1988.

[30] Pryor, F. Income Distribution and Economic Development in Malawi. WDP 36. World Bank, 1988.

[31] Mussa, R. and Pauw, K. "Poverty in Malawi: Current status and knowledge Gaps." *Policy Note* 9, 2011. *See also* Mussa. R. Poverty and Inequality in Malawi: Trends, Prospects, and Policy Simulations. MPRA no. 75979, 2017 https://mpra.ub.unimuenchen.de/75979. Mangani, R., Jayne, T., Hazel, P., Muyanga, M., Chimatiro, S., Burke, W. & Johnson, M. Agricultural Transformation in Malawi: Call to Action. *MWAPATA Institute Working Paper N*o. 20/01, 2020.

Broadly speaking, the incidences of poverty increased by 0.3% per year on average from 2003 to 2008 and 0.9% from 2008 to 2015, reaching as high as 75.7% in 2015.[32] Poverty has been getting worse over the years. The poverty rate in Malawi in 2015, Mangani et al. lament, was the highest of all Southern African countries.[33] Studies have shown that one-fifth (20%) of the population in Malawi lives in extreme poverty, while more than half (50%) live in moderate poverty.[34]

In order to fill the inequality gap and indeed to significantly reduce the level of poverty in Malawi, the government has, in various regimes, used a range of poverty reduction strategies. These strategies, however, have failed to make any significant positive impact. For example, in the 2005/2006 growing season, during the Bingu wa Mutharika regime, the government implemented an agricultural input subsidies programme, christened the Farm Input Subsidy Programme (FISP), in order to remedy the severe food supply shortages in the country.[35] The prime aim of the programme was to provide low-cost maize seed and fertilizer to poor peasants and small-scale farmers. While FISP was praised at a local and regional level for boosting maize production in that growing season alone, it reportedly widened further the inequality gap and accelerated poverty levels among the rural poor. Mussa and Pauw observed that the net consumers of the maize that was produced from 2004 to 2010, that is, urban households, benefited from the low prices arising from an increase in maize production, while it disadvantaged the rural majority, especially the net producers, because of the uncharacteristically low prices of maize.[36] It is, therefore, safe to argue that the inequality gap between the rich and the poor in Malawi has been, in fact, exacerbated by the poor and unequal distribution of scarce resources in government interventions such as the agricultural input subsidies and social cash transfers that have not benefited the majority of the poorest of the population. Agricultural input subsidies face serious procurement and distribution challenges, leading to the precarity of the ultra-poor who continue their human-made chronic dependency on such government handouts as FISP.

What is of note here is the grim reality that, in spite of various regimes' focus on them, agricultural input subsidies have not succeeded in solving the problem of hunger. Most of the beneficiaries of agricultural input subsidies, the ultra-poor, end up selling their inputs (sometimes they sell input entitlements, i.e., coupons) to vendors

---

[32] Mussa, R. and Pauw, K. "Poverty in Malawi: Current status and knowledge Gaps." *Policy Note* 9, 2011.

[33] Mangani, R., Jayne, T., Hazel, P., Muyanga, M., Chimatiro, S., Burke, W. & Johnson, M. Agricultural Transformation in Malawi: Call to Action. *MWAPATA Institute Working Paper N*o. 20/01, 2020.

[34] Mangani, R., Jayne, T., Hazel, P., Muyanga, M., Chimatiro, S., Burke, W. & Johnson, M. Agricultural Transformation in Malawi: Call to Action. *MWAPATA Institute Working Paper N*o. 20/01, 2020. See also World Bank. United Nations Population Division, World Population Prospects: 2022 revision, 2020.

[35] Mussa, R. and Pauw, K. "Poverty in Malawi: Current status and knowledge Gaps." *Policy Note* 9, 2011.

[36] Mussa, R. and Pauw, K. "Poverty in Malawi: Current status and knowledge Gaps." *Policy Note* 9, 2011.

or to slightly less poor farmers, and even to the well-off who can otherwise afford to buy the agricultural inputs at market prices.[37] This wanton sale of agricultural inputs and input entitlements secures them a temporary supply of food, just for a few days; in the long term, however, it worsens the poverty levels as they are later expected to buy staple food at exorbitant prices from vendors. In addition to inclusion errors, such as being of benefit to smallholder farmers who are not supposed to receive them, the sale of coupons, fraud, and corruption have also had a negative effect on the agricultural input subsidies programmes from meeting their intended objectives,[38] further deepening the poverty levels among the already ultra-poor population. Between 2005 and 2020, poverty levels among the rural poor have become much higher despite the continued enrolment in the agricultural input subsidies programmes.[39] In the immediate past growing season at the time of writing (that is, the 2022/23 growing season), crops, especially maize, did not do well due to a wide range of factors, ranging from the massive crop destruction by flash floods that came with Cyclone Freddy and the anomalies associated with the year's agricultural input subsidies programme—the Affordable Inputs Programme (AIP). The government delayed fertilizer procurement, and most ultra-poor did not benefit from the same, which resulted in poor harvests in 2023.[40] This has resulted in a scarcity of maize and skyrocketing food prices in Malawi. At the time of writing, a 50 kg bag of maize was going for MK40,000, a price which was too high for the ordinary Malawian, putting the majority of the population at a high risk of hunger. The AIP distribution and logistical challenges, coupled with the dependency syndrome it creates among the ultra-poor segment of the population, severely worsened hunger and poverty in Malawi then.

In addition to the agricultural input subsidies, another government intervention intended to narrow the inequality gap between the rich and the poor is the Social Cash Transfer (hereinafter SCT). While the SCT was initially designed to reduce poverty and hunger and increase school enrolment and attendance, it has not actually achieved its intended objectives. Malawi's social protection policy framework is widely observed to fall short of the country's current and future socio-economic needs. SCTs' effectiveness becomes compromised by shortfalls such as *ad hoc* and uncoordinated implementation, inadequate funding, inherent programme design weaknesses, and capacity constraints.[41] The amounts dispensed under the Social

[37] Chirwa, E.W., and Dorward, A. *Agricultural input subsidies: The recent Malawi experience*. Oxford University Press, 2013.

[38] Tambulasi, R. I. C. "The public sector corruption and organised crime nexus: The case of the fertiliser subsidy programme in Malawi." *African Security Review 18*, (2009): 19–31.

[39] Chirwa, E.W, and Dorward, A. *Agricultural input subsidies: The recent Malawi experience*. Oxford University Press, 2013. National Statistical Office. Malawi Poverty Report of 2020. Zomba, 2021.

[40] *The Nation*. 2022/23AIP: Under wraps and muddled up, 2022.

[41] Miller, C.M., Tsoka, M., and Reichert, K. Impact Evaluation Report External Evaluation of the Mchinji Social Cash Transfer Pilot, a report prepared by the Center for International Health and Development (CIHD) Boston University for the Government of Malawi, USAID and UNICEF Malawi, 2008.

Cash Transfer programme, for instance, are too small to make any noteworthy positive impact on the intended beneficiaries, the ultra-poor. SCTs reach few of the ultra-poor and the most vulnerable people in Malawi, the majority of whom are found in the rural parts of the country.[42] For example, in 2015, the Social Cash Transfer programme met only half of its 10% target. It should be noted that Malawi's public spending on social protection is extremely low by worldwide standards, and, indeed, even lower than the average for Africa's low-income countries.[43]

In some respects, SCTs have widened the inequality gap and worsened the poverty levels among the ultra-poor. Other studies have established that the inaccurate targeting of beneficiaries fails to take into consideration small economic differences among the overall percentages of the population who are ultra-poor: "No wonder …cash transfers have been divisive because there is very little difference between beneficiaries and the rest of their communities in well-being, access to assets and income and real consumption."[44] Porzany and O'Brien, for instance, aptly observed that while "beneficiaries gained access to networks requiring financial contributions, communities often excluded them from other social programmes for equity reasons, even though this is not within the policy."[45] As a result of jealousy, these people's personal ties are again adversely affected.[46] In general, therefore, Social Cash Transfers are bedevilled with serious logistical and distribution challenges to the extent of bypassing most of the ultra-poor. SCTs aside, even land (re-)allocation has witnessed estates—an exclusive privilege of the upper middle class, i.e. the affluent—benefiting at the expense of the ultra-poor.

10.5 Discussion

Malawi's population boom, coupled with the high levels of poverty as illustrated by the widening inequality gap, is a matter of great politico-ethical concern for policymakers in both the public (governmental) and the private (non-governmental) sectors. We have seen that government interventions such as agricultural input subsidies and Social Cash Transfers have failed dismally to narrow the inequality gap between

[42] Kalebe-Nyamongo, C. and Marquette, H. "Elite Attitudes towards Cash Transfers and the Poor in Malawi," 2014. http://publications.dlprog.org/EliteAttitudesCTs.

[43] Chirwa, E.W, and Dorward, A. *Agricultural input subsidies: The recent Malawi experience*. Oxford University Press, 2013.

[44] Jimu, I. and Msilimba, G. "Targeting Practices and Biases in Social Cash Transfers." *Africa Development 43, no.* 2(2018): 65–84.

[45] Porzany, P. and O'Brien, C. The impacts of Malawi's social cash transfer programme on community dynamics. *The International Policy Centre for Inclusive Growth* 276, 2015, p. 1.

[46] Miller, C.M., Tsoka, M., and Reichert, K. Impact Evaluation Report External Evaluation of the Mchinji Social Cash Transfer Pilot, a report prepared by the Center for International Health and Development (CIHD) Boston University for the Government of Malawi, USAID and UNICEF Malawi, 2008.

the rich and the poor. In such a context, both individual persons and communities cannot claim and enjoy rights, most especially socio-economic rights. A good-quality life, we argue, is a socio-economic right. The inequality gap between the rich and the poor is indicative of selfishness on the part of the rich, who are as individualistic as they are narrowly focused on their self-realization, and hence are far from 'other-regarding' (virtues or duties). Ubuntu philosophy can be used as an antidote to Malawi's widening inequality gap because of its 'other-regarding virtues, such as sharing with others what one possesses. The duty of beneficence is at the heart of Ubuntu philosophy. For Ubuntu philosophers such as Motsamai Molefe, the personhood of Malawi's rich is in question because:

> The idea of personhood entails a purely other-regarding morality of duties; it is these duties that take priority over rights. Rights are considered to be secondary because they will have a tendency to clash with the fundamental moral-social goal posited by a morality of duties for securing the well-being of all human beings, entailed by the idea of personhood.[47]

It follows from the above-cited quotation that the rich have other-regarding duties towards the poor—for example, the duty of beneficence—so that the poor can also lead a meaningful life as persons. Molefe thinks of a meaningful life roughly as a life that matters or is morally significant:

> We tend to care about lives that (morally) matter or are (morally) significant because we deem them to be objectively fulfilling. The objective status of meaningful lives being fulfilling renders them choice-worthy, i.e., the kinds of positive lives that anyone ought to aspire to live.[48]

It is obvious that no one can choose to lead a life of poverty, destitution, or deprivation because such a life is not meaningful inasmuch as it is not choice-worthy. The life of the poor is not objectively fulfilling.

The Western individualism that Africa has inherited from its colonial past is to blame for the widening inequality gap between the rich and the poor in Malawi and in other Southern African countries. As a result of this colonial legacy, today we find that African ethics is disappointingly and disproportionately tainted with Western individualism. Molefe identifies individualism in modern African moral cultures, thus:

> Individualism is a central feature of African ethics. Individualism refers to those moral theories that ultimately ground morality on some properties intrinsic to an individual, like welfare, soul, and rights.[49]

Another component of African ethics is that of African communitarianism, which means that the community has primacy over the individual. Ubuntu philosophers

---

[47] Molefe, M. "Individualism in African moral cultures." *Cultura: International Journal of Philosophy of Culture and Axiology 14, no. 2* (2017): 49-68. *See also* Molefe, M. "Personhood and Rights in African tradition." *Politikon 45, no.* 2(2018): 217–231.

[48] Molefe, M. "Personhood and a meaningful life in African Philosophy." *South African Journal of Philosophy 39, no.* 2(2020):194–207.

[49] Molefe, M. "Individualism in African moral cultures." *Cultura: International Journal of Philosophy of Culture and Axiology 14, no.* 2(2017): 49–68.

such as Mogobe Ramose, Desmond Tutu, Thaddeus Metz, Grivas Kayange, and Motsamai Molefe, among others, emphasize this component in their articulations within African Philosophy. For example, Molefe and Magam ask what Ubuntu can offer in the light of post-colonial challenges that Africa faces:

> Ubuntu is attractive to its proponents because it is purported to have two features: on the one hand, it offers an African moral perspective on issues; and, on the other, it is believed to offer an under-explored morally plausible vision of a good society…What makes matters difficult, however, is that there are competing conceptions of Ubuntu in the literature, and they all offer differing visions of what might count as a robust conception of a society.[50]

A person who is trapped in poverty cannot possibly self-realize. The personhood of such a poor, destitute, or deprived individual is grossly compromised, even in the Ubuntu philosophy, where self-realization is crucial. On self-realization, Molefe and Magam argue as follows:

> The self-realization approach to Ubuntu takes moral individualism as its moral point of departure. That is, this interpretation of Ubuntu ultimately values individuals realising their own true humanity as the goal of morality. This way of understanding Ubuntu tends to be captured in terms of the normative idea of personhood in the literature. This idea of personhood is taken from the maxim that captures the essence of Ubuntu as a moral theory—a "person is a person through others."[51]

The maxim 'a person is a person through others' will not have any purchase in a population where resources are very scarce and too few to be distributed fairly, where there is an ever-widening inequality gap between the rich and the poor, and where the majority of persons are trapped in poverty, hunger, disease, and ignorance (or lack of formal education).

Even the idea of dignity cannot be associated with a population like that of Malawi in which the majority of the people are trapped in poverty, destitution, or deprivation. For Molefe, "Dignity is used usually to refer to a quality of beings of superlative value, those towards whom we take ourselves to have maximum duties of respect."[52]

For Molefe, "a plausible conception of dignity ought to essentially involve how the agent positively uses these (ontological) capacities to live, informed by principles that enrich the community…, the capacity for, and manifestation of, self-respect, and respect for, and responsibility to, others."[53] However, Molefe objects to this grounding of dignity in personhood because, for him, "Dignity is not the kind

---

[50] Molefe, M. and Magam, N. "What can Ubuntu do? A reflection on African moral theory in light of post-colonial challenges." *Politikon 46*, no. 3(2019): 311–325.

[51] Molefe, M. and Magam, N. "What can Ubuntu do? A reflection on African moral theory in light of post-colonial challenges." *Politikon 46*, no. 3(2019): 311–325.

[52] Molefe, M. *An African Philosophy of Personhood, Morality and Politics*. Palgrave Macmillan, 2019, p. 117.

[53] Molefe, M. *An African Philosophy of Personhood, Morality and Politics*. Palgrave Macmillan, 2019, p. 119.

of property that varies among agents relative to the conduct of agents. Dignity is a function of merely being *human,* and all human beings have it in equal measure."[54]

By implication, we derive the maxim that all human beings are dignified (or have dignity) simply because they all deserve dignity in equal measure. Human beings have human dignity without qualification. This maxim is egalitarian in nature and scope. However, the egalitarian conception of dignity becomes severely tested and compromised in a socio-economically stratified society such as Malawi, where there is a yawning inequality gap between the rich and the poor. Where is the dignity for a major segment of the population that goes hungry for days and cannot afford one out of the regular three meals per day? Where is dignity for a major segment of the population of people with no secure and safe shelter? Where is the dignity for a major segment of the population of human beings who do not have decent clothes? Where is dignity for a major segment of the population of human beings who have no access to family planning methods and sexual and reproductive healthcare? These questions are not rhetorical; rather, they point to a severe lack of resources that would enable a meaningful and objectively fulfilling life for the majority of the Malawian population, which is rapidly and alarmingly growing, leading to what we characterize as *population boom.* Since Malawi's rich are not altruistic, these non-rhetorical questions crucially point to a lack of a political economy for (re-)distributing fairly the resources that are now only in the hands of the super-rich few. By implication, that is, drawing on Molefe, it can be asserted without fear of contradiction that Malawi's wealthy few are starving and impoverishing Malawi's majority poor.

Staying within Ubuntu philosophy, the ever-widening inequality gap between the rich and the poor in Malawi can further be elucidated by looking at it in terms of public morality, dignity, and human rights. For example, Thaddeus Metz thinks that, if suitably interpreted, Ubuntu can serve as a ground of public morality. He defends the view that "even if various construals of Ubuntu have been vague, collectivist and anachronistic, it can be interpreted in a promising way."[55] His "favoured moral theory is that actions are right, or confer Ubuntu (humanness) on a person, insofar as they prize communal relationships, ones in which people identify with each other or share a way of life, and exhibit solidarity toward one another, or care about each other's quality of life."[56] Metz extends his relational conception of Ubuntu to human rights in the entirety of Southern Africa. Thus, for Metz:

> Such a principle has a Southern African pedigree, provides a new and attractive account of morality, which is grounded on the value of friendship, and suggests a novel, companion conception of human dignity with which to account for human rights. According to this

[54] Molefe, M. *An African Philosophy of Personhood, Morality and Politics*. Palgrave Macmillan, 2019, p. 117.

[55] Metz, T. "Ubuntu as a moral theory and human rights in South Africa." *African Human Rights Law Journal 11*, no. 2 (2011): 532–552.

[56] Metz, T. "Ubuntu as a moral theory and human rights in South Africa." *African Human Rights Law Journal 11*, no. 2 (2011): 532–552.

> conception, typical human beings have dignity by virtue of their capacity for community or friendliness, where human rights violations are egregious failures to respect this capacity.[57]

Metz claims to be capable of deriving human rights from "salient elements of Ubuntu." He argues as follows:

> I posit a single basic good, that of human dignity *qua* capacity for communal relationship, and I maintain that it is to be respected. My key claim is that part of what it means to treat someone as special by virtue of her capacity for communal or friendly relationship is to commune with or be friendly towards her (at least when she has not misused that capacity), and that what it means to be friendly towards a person is to avoid subordinating her, amongst other conditions such as avoiding harming her and acting consequent to negative attitudes such as cruelty.[58]

From the above-cited quotation, it is implied that because of human dignity intrinsic to the individual, a human person has a human right in communal or friendly relationships in a traditional Sub-Saharan African context. This human right entails that others do not harm or subordinate her in any way.

Metz elaborates on the communitarian conception of human rights in Africa. He proffers a promising way to reconcile a communitarian framework with a robust prizing of human rights—that is, despite the individualistic nature of human rights. Metz (2014b, 132) avers that "the idea of individual rights in fact comports well with prominent African judgments of human rights." For Metz,

> A human right is of course a particular kind of right, one that has three properties, at least one which is lacking in other kinds of rights. First, inherent to the concept of a human right is that its bearer is an individual, and, specifically, least controversially, a human person…Second, a human right is essentially such that others have a stringent duty to the individual because she has some quality of characteristic human beings. Roughly, one has a human right in virtue of one's humanity…Third, it is part of the meaning of the phrase 'human rights' to be speaking of duties that are so important as to warrant recognition by at least legal institutions.[59]

From the above-cited quotation, Metz is implying that the bearer of a human right is, at the very least, a human person and that other human persons have stringent duties to her, one of which is the duty not to harm her, because she partakes of humanity or simply because she is human. Additionally, such a human rights bearer is recognized by legal institutions.

---

[57] Metz, T. "Ubuntu as a moral theory and human rights in South Africa." *African Human Rights Law Journal 11*, no. 2 (2011): 532–552.

[58] Metz, T. "African Values and human rights as two sides of the same coin. A reply to Oyowe." *African Human Rights Law Journal 14*, no. 1 (2014a): 306–321.

[59] Metz, T. African values, human rights and group rights: a philosophical foundation for the Banjul Charter, (2014b) 131–152. In: O. Onanzi (ed.) *African Legal Theory and Contemporary Problems: Critical Essays. Ius Gentium:Comparative Perspectives on Law and Justice.* Springer, 2014.

10.6 Conclusion

There is a hugely worrisome mismatch between the population of Malawi and the available resources on the ground to cater to the booming population. The alarming increase in the population, which is met with scarce resources such as food, has rendered the majority segment of the population vulnerable to hunger and famine every growing season. There is a general condition of chronic hunger and abject poverty across the country. For example, in the 2023–2024 growing season alone, over four million people were reportedly facing hunger. The ever-widening inequality gap between the rich and the poor continues to exert immense pressure on the government to intervene using the welfarist tools of agricultural input subsidies and social cash transfers, albeit to date with little to no impact on hunger and poverty. Non-governmental organisations such as the UN World Food Programme also try their best to complement the government's efforts by providing relief maize, mostly among the rural poor. Poverty levels remain disproportionately high in Malawi's population despite government welfarist interventions. Government welfarist interventions have dismally failed to ease the pressure that the ever-growing population exerts on scarce resources. To make matters worse, Malawi's rich have not been moved ethically to complement the government's efforts in combating hunger and poverty among the majority segment of the booming population.

For Ubuntu philosophers such as Motsamai Molefe and Thaddeus Metz, the personhood of Malawi's rich can be argued to be seriously in question. We argue that it is largely due to the unethical conduct of the rich that the poor are not in a position to lead a quality life in Malawi, being severely lacking in basic needs such as food, shelter, clothing, and potable water. Malawi's rich, most of whom conduct themselves unethically vis-à-vis their immediate communities, do not exhibit 'other-regarding' morality of duties or a public morality of communal relationality and friendliness. Malawi's rich dismally fail to relate to, and are not friendly to, their poor neighbors. According to John Makina (as cited in Mussa and Masanjala 2015, 5), economic inequality has worsened significantly in Malawi in recent years. In 2004, the richest 10% of Malawians consumed 22 times more than the poorest 10%. By 2011, this had risen to the richest 10% spending 34 times more than the poorest 10%. In a typical African communitarian and public morality, duties take priority or precedence over rights, which are grounded in individualism. Of course, individuality is another component of the referred-to African communitarian and public morality. As Metz has forcefully and repeatedly argued in various fora, the key point to emphasize here is that in a traditional Sub-Saharan African community, the bearer of a human right is, at the very least, a human person, and that other human persons have stringent duties to her. We herein extend this African communitarian and public morality and henceforth argue that if the rich callously leave the poor to starve (to death), then the rich are directly or indirectly inflicting harm on the poor and hence the rich are morally blameworthy. It is the communal and public moral responsibility of the government, the rich, and non-governmental organizations to feed the country's booming population. Bad governance, especially rampant public corruption, aggravates the condition of the poor in Malawi. For example, under the Joyce Banda presidency (2012–2014), in 2013, billions of Malawi Kwacha (over K23 billion) in government coffers vanished into thin air due to a porous financial accounting system. The local news media dubbed this public financial scandal "Cashgate." Suppliers of goods and providers of services to the government at Capital Hill were fraudulently paid huge sums of money without supplying or providing anything to the government. Most of the culprits of Cashgate are yet to be properly prosecuted and punished. The few culprits that were apprehended and eventually convicted through the courts served very light prison sentences and almost all of them are already out of prison to date.

Ultimately, the Malawi Government and relevant non-governmental organizations ought to conscientize and strongly encourage the poor to adopt and follow family planning methods so as to keep the population growth rate under control since economically (in economic theory), in spite of government's and nongovernmental organizations' efforts, policy interventions and strategies, resources such as food and money will remain scarce in Malawi.

References

Chirwa, E. W., and A. Dorward. 2013. *Agricultural input subsidies: The recent Malawi experience*. Oxford University Press.

Cornia, G., and B. Martorano. 2015. *Building the IID–SSA inequality dataset and the seven sins of inequality measurement in SSA*. Mimeo, UNDP Project on SSA Inequality.

Jimu, I., and G. Msilimba. 2018. Targeting practices and biases in social cash transfers. *Africa Development* 43 (2): 65–84.

Kalebe-Nyamongo, C., and H. Marquette. 2014. *Elite attitudes towards cash transfers and the poor in Malawi*. http://publications.dlprog.org/EliteAttitudesCTs

Kalinga, Owen. 1979. Trade, the Kyungus, and the emergence of the Ngonde Kingdom of Malawi. *The International Journal of African Historical Studies* 12 (1): 17–39.

Kalinga, O., and J. Pike. 2000. A pre-colonial history of Malawi. *The Nyasaland Journal* 18 (1): 22–54.

Malawi Population Census. 1977. *Analytical report Volume 2 Infertility, Mortality and Population Projections 1901-2000*.

Mangani, R., T. Jayne, P. Hazel, M. Muyanga, S. Chimatiro, W. Burke, and M. Johnson. 2020. *Agricultural transformation in Malawi: Call to action*. MwAPATA Institute Working Paper No. 20/01.

Mbewe, S. M., and F. Kholowa. 2010. The Malawi education system and reproduction of social inequalities: Implications of education for sustainable development. *Malawi Journal of Education and Development 3*:63–74.

Metz, T. 2011. Ubuntu as a moral theory and human rights in South Africa. *African Human Rights Law Journal* 11 (2): 532–552.

Metz, T. 2014a. African Values and human rights as two sides of the same coin. A reply to Oyowe. *African Human Rights Law Journal* 14 (1): 306–321.

Metz, T. 2014b. African values, human rights and group rights: A philosophical foundation for the Banjul Charter. In *African legal theory and contemporary problems: Critical essays. Ius Gentium: Comparative perspectives on law and justice*, ed. O. Onanzi, 131–152. Springer.

Miller, C. M., M. Tsoka, and K. Reichert. 2008. *Impact Evaluation Report External Evaluation of the Mchinji Social Cash Transfer Pilot, a report prepared by the Center for International*

Health and Development (CIHD). Boston University for the Government of Malawi, USAID and UNICEF Malawi.

Molefe, M. 2017. Individualism in African moral cultures. *Cultural: International Journal of Philosophy of Culture and Axiology* 14 (2): 49–68.

Molefe, M. 2018. Personhood and rights in African tradition. *Politikon* 45 (2): 217–231.

Molefe, M. 2019. *An African philosophy of personhood, morality and politics*. Palgrave Macmillan.

Molefe, M. 2020. Personhood and a meaningful life in African philosophy. *South African Journal of Philosophy* 39 (2): 194–207.

Molefe, M., and N. Magam. 2019. What can Ubuntu do? A reflection on African moral theory in light of post-colonial challenges. *Politikon* 46 (3): 311–325.

Mussa, R. 2017. *Poverty and inequality in Malawi: Trends, prospects, and policy simulations*. MPRA no. 75979. https://mpra.ub.uni-muenchen.de/75979/

Mussa, R., and W. Masanjala. 2015. *A dangerous divide: The state of inequality in Malawi*. Oxfam.

Mussa, R., and K. Pauw. 2011. Poverty in Malawi: Current status and knowledge gaps. *Policy Note* 9.

Mwakilama, S. 2018. Poverty and inequality and the quest for inclusivity: A case study of Malawi at 54 years of independence. *American Journal of Social Sciences* 6 (4): 65–72.

National Statistical Office. 2021. *Malawi poverty report of 2020*. Zomba.

Porzany, P., and C. O'Brien. 2015. The impacts of Malawi's social cash transfer programme on community dynamics. *The International Policy Centre for Inclusive Growth* 276.

Pryor, F. 1988. *Income distribution and economic development in Malawi*. WDP 36. World Bank.

Sindima, H. 2002. *Malawi's First Republic: An economic and political analysis*. University Press of America.

Tambulasi, R. I. C. 2009. The public sector corruption and organised crime nexus: The case of the fertiliser subsidy programme in Malawi. *African Security Review 18*:19–31.

The Nation. 2022/23AIP: Under wraps and muddled up, 2022.

Thompson, T. 1981. The origins, migration and settlement of the Northern Ngoni. *The Society of Malawi Journal* 34 (1): 6–35.

World Bank. 2020. *United Nations Population Division, World Population Prospects: 2022 revision*.

Part IV
Case Studies and Contemporary Issues

Chapter 11
An Ethical Exploration of Young People's Perception of Human Population Growth in Contemporary South Africa

Nothando Happy-Girl Shandu and Beatrice Okyere-Manu

Abstract In recent times, the issue of population growth has come to the fore. Globally, ongoing conferences, debates, and discussions have been held to explore strategies to respond to the ever-growing population. This is because of its effect on the environment, resource management, and future generations. The African population is no different. Since time immemorial, the African population has continued to grow exponentially, and the result is the current challenges that the continent is facing. With this background, this chapter ethically explores the attitudes, perceptions, and beliefs of young people in South Africa concerning the ongoing discourse on human population growth. It critically reviews existing literature on fertility and population growth in Africa using a qualitative research methodology.

Keywords Limited communitarianism · Traditional communities · Future generations · Urbanization · Overexploitation

Through the ethical theory of limited communitarianism, the study exposes the values, concerns, and moral considerations that inform young people's views on African population growth. The findings are expected to contribute to a detailed understanding of the ethical implications of the changing demographic trends, offering valuable insights for policymakers and the traditional communities involved in addressing population-related challenges.

N. H.-G. Shandu (✉)
Philosophy (Applied Ethics), University of KwaZulu-Natal, Pietermaritzburg, South Africa
e-mail: shandun@ukzn.ac.za

B. Okyere-Manu
Department of Philosophy and Ethics Studies, University of KwaZulu-Natal, Pietermaritzburg, KwaZulu-Natal, South Africa
e-mail: okyere-manv@ukzn.ac.za

B. Okyere-Manu, P. Mwipikeni (eds.), *An African Approach to Population Ethics*, Advancing Global Bioethics 23,
https://doi.org/10.1007/978-3-032-24464-2_11

11.1 Introduction

Over the past two centuries, the world's human population has grown exponentially. According to Christine Kaneda, Asia is currently the world's most populous continent (4.81 billion), followed by Africa with 1.52 billion people.[1] Rahim Sadigov calculates that Asia makes up nearly 60% of the world's population, while Africa makes up approximately 18%, leaving the remaining 22% of the world's population to other regions.[2]

According to a forecast by Statista Research Department, Africa's total population could reach nearly 2.5 billion by 2050. In 2023, the continent had around 1.36 billion inhabitants, with Nigeria, Ethiopia, and Egypt as the most populous countries. The discrepancy arises because the figures are drawn from different sources and years, with some representing current estimates and others based on projections. In the coming years, Africa will experience significant population growth and close the gap significantly with the Asian population by 2100.[3] This population growth results from an increase in the birth rate and a decrease in the mortality rate. Overpopulation occurs when a high density of people leads to environmental degradation, reduced living standards, famine, or conflict.[4] It is a term whuch is commonly used to describe the relationship between the human population and the environment. Due to the rapid increase in the global population over several decades, overpopulation has emerged as a major environmental concern across the globe. As noted above, Asia constitutes almost 60% of the world's population; however, it is Africa which is often considered overpopulated because it has the highest population growth rate among all continents. The total fertility rate of 4.3 births per woman in Africa is 2.1 higher than that for the rest of the world.[5] Therefore, despite the ongoing efforts to decrease its fertility rate, the population of Africa, specifically the sub-Saharan region, is predicted to double by the year 2050, resulting in overpopulation. The growth of the human population is expected to have adverse effects on the continent's already limited resources and urban infrastructure, including transportation, housing, and public services. Sadigov argues that overpopulation, particularly in those regions heavily reliant on agriculture and forestry, exacerbates the pressure on urban infrastructure and intensifies environmental degradation and cli-

[1] Kaneda, C. P. (2023). World Population data-sheet-boolet. Climate Vulnerability and Resilience, 1, 1–25. 12.

[2] Sadigov, R. (2022). Rapid Growth of the World Population and Its Socioeconomic Results. Scientific World Journal, 2022 (1930). https://doi.org/10.1155/2022/8110229. 2.

[3] Statista Research Department, Mar 22, 2024. Forecast of the total population of Africa 2020–2050. https://www.statista.com/statistics/1224205/forecast-of-the-total-population-of-africa/. Accessed date: 23–08–2024.

[4] John Bongaarts, 'Human Population Growth and the Demographic Transition', *Philosophical Transactions of the Royal Society B: Biological Sciences*, 364.1532 (2009), 2985–90 <https://doi.org/10.1098/rstb.2009.0137>.

[5] United Nations, *Department of Economic and Social Affairs, Population Division (2022), World Population Prospects 2022: Summary of Results.*, 2022 <UN DESA/POP/2022/TR/NO. 3>.

mate change.[6] Amadu Kaba states that several demographic, social, and economic factors have contributed to Africa's rapid population growth.[7] These include high fertility rates, low usage of contraceptives, a decline in infant mortality, and a reduction in overall death rates. There are, however, some considerable demographic variations across the five regions of Africa.[8] In 2023, South Africa's fertility rate was 2.3 births per woman, a 0.9 decline from 2022.[9] Although this marks a steady decline in fertility, South Africa continues to grapple with a relatively high population growth rate of approximately 1.3 higher than in 2023.[10] This problem is anticipated to persist, given the significant prevalence of teenage pregnancies within the country. As of 2022, the age group comprising children between 10 and 19 years, totaling over 10.8 million individuals, represents the largest portion of South Africa's population. Other factors contributing to population growth include increased life expectancy and decreased infant mortality, which reflect advances in medicine, public health, and socioeconomic development.

The current chapter investigates young people's attitudes, perceptions, and beliefs regarding the ongoing discourse on human population growth in South Africa. It aims to expose the perspectives of the younger generation in the context of contemporary Africa. This will be accomplished using the ethical theory of limited communitarianism and qualitative research methods to uncover the values, concerns, and moral considerations that shape young people's perspectives with regard to population growth. The findings are expected to contribute to understanding the ethical dimensions inherent in changing demographic trends, offering valuable insights for policymakers and the traditional communities involved in addressing population-related challenges. The chapter is divided into four sections. This introductory section has presented the context and the current issues related to human population growth in Africa, specifically focusing on South Africa. Next, the chapter explains the methodology employed in the study. This is followed by a discussion on the ethical reflections of the findings gathered from the youth. Finally, the conclusion will be presented.

---

[6] Sadigov, Rapid Growth, 4.

[7] Kaba, A. J. (2020). Explaining Africa's Rapid Population Growth, 1950 to 2020: Trends, Factors, Implications, and Recommendations. Sociology Mind, 10(04), 226–268. https://doi.org/10.4236/sm.2020.104015

[8] Amadu Jacky Kaba, 'Explaining Africa's Rapid Population Growth, 1950 to 2020: Trends, Factors, Implications, and Recommendations', *Sociology Mind*, 10.04 (2020), 226–68 <https://doi.org/10.4236/sm.2020.104015>.

[9] Christine Power Kaneda, 'World Population Data -Sheet-Boolet', *Climate Vulnerability and Resilence*, 1 (2023), 1–25.

[10] United Nations, *Department of Economic and Social Affairs, Population Division (2024), World Population Prospects 2024: Methodology of the United Nations Population Estimates and Projections*, 2024 <UN DESA/POP/2024/DC/NO. 10, July 2024>.

11.2 Methodology

This study is based on responses from postgraduate students over the past five years in the course *Environmental Ethics: An African Perspective* offered at the Honours level in the School of Religion, Philosophy, and Classics at the University of KwaZulu-Natal, South Africa. Every year, an average of eight students register for this module. Thus, from the introduction of the course in 2016 right up to 2023, about 50 students have registered for the course. These students are often between the ages of 20 and 35 and are come from diverse ethnic, religious, and cultural backgrounds. Of the 50 students, 15 were males, and 35 were females. Among the numerous themes in this module are the subjection of population growth and its effect on the environment. The purpose has been to examine the perceptions of Africa's population and the environmental crisis. The class discussions on this theme have always been interesting and engaging. Among the questions that have sparked long debates and discussions are: Is Africa overpopulated as commonly believed? If so, what are the causes and consequences of this perception? How does Africa view population growth? How do we ensure people have a decent life? Are we obligated to create more people afflicted by low quality of life or to have fewer people with a high quality of life? What steps can Africa take to secure a sustainable future for current and future generations?

These questions have always sparked exciting and heated debates among the students in the classroom. The responses, and the issues raised over the years, informed the decision to expose the young people's perspective on the growth of the African human population. Over the 7 years, the responses and discussions have been recorded with permission from the students. Pseudonyms will be used to protect the identities of the students. The data will be analyzed using content analysis, which presents the data in either words or themes to make it possible to draw some interpretations of the result.[11]

11.3 The Findings

The following section presents a compilation of the responses to the above questions over five years. This will be presented in terms of broad themes.

[11] Bengtsson Marriette. 2016. How to plan and perform qualitative study using content analysis. Nursing Plus Open. 2, (8–14). 10.

11.4 Africa and Overpopulation

One of the recurring themes in our class discussions over the years has been that Africa is not overpopulated. The ethics students believe that the saying that to state that the entire continent is overpopulated would be an exaggeration. They agree that the African population is growing because of urbanization, not high fertility. As a result of poverty and the lack of infrastructure, most rural people migrate to cities, increasing the urban population. This does not mean, however, that the continent is overpopulated because the rural areas are mostly vacant. A student in the 2019 group said:

I don't believe in the argument that Africa is overpopulated; look at our country, South Africa. There are large stretches of land that are not being occupied. The apartheid government designated unproductive and small portions of land to black people. So, when you look at the black communities, you would think we are over-populated, but we are not. If we spread ourselves, one will see that there will even be excess unoccupied lands.

The 2023 cohort class discussions echoed a similar sentiment. A male student said that:

> Countries in Africa are not overpopulated like other countries. We have lands that are not occupied, and in most rural communities, the homesteads are scattered. In some cases, the distance between the homesteads is one kilometer. There are only a few people in most villages. How can we say that such a continent is overpopulated?

From the above, it is clear that even though the fertility rate is high, the continent is not overpopulated, particularly when comparing the vastness of the continent's land to the number of people, as indicated in the introduction of this chapter. An Emerging Markets Forum's document reveals that Africa is "the second largest population and also the second largest land mass (after Asia) of all regions."[12] The document suggests that even though the continent's population is large, there is sufficient land to accommodate its people.

Even though most of the students on the course believe that Africa is not over-populated, a handful also believe that the popular assertion that Africa is overpopulated is that the continent's cities are overcrowded because of urbanization. It was unanimously agreed that this is because of the lack of resources in small towns and rural areas. As a result, young men and women flock to the nearby cities for employment and survival, placing a strain on the resources in urban areas.

Over the years, some other students have agreed with the popular assertion that Africa is overpopulated. For instance, in the 2018 cohort, a female student argued that she believes Africa is overpopulated: "if Africa is not overpopulated, then why is poverty everywhere? People keep giving birth, and the government cannot afford to support us in schools; look at our hospitals; there are people everywhere. What do

[12] Emerging Markets Forum. Africa 2050: Realising the Continent's full potential https://www.jica.go.jp/Resource/jica-ri/publication/booksandreports/jrft3q00000029i2-att/Africa_2050_Overview_final_29_May.pdf. ix.

you call that? Isn't it overpopulation?" We have also had students who believe that the continent is neither overpopulated nor underpopulated, but that it is broadly average, with a reasonable distribution of population.

11.5 Personal Preference

On the theme of their personal preferences regarding the number of children they anticipate, only a few say that they are willing to have two children. Most of the students' preferences are for only one child. The following are the reasons why they wouldn't have many children:

> I suffered while growing up, so I don't want my children to suffer. We were 8 children at home, and food and other resources were scarce.

A female from the 2020 cohort shared an experience that has made her reconsider her preference for more children:

> I have seen families and our neighbors with large numbers struggle to look after the children in their homes. Sometimes, some children come to our home to beg for food. I know a family suffering because there are many mouths but not enough income. The whole household depends on the grandmother's pension, and as a result, some of the children have become street children.

The responses above reflect how the students' experiences in terms of what they have gone through in their own families and what they have seen in their communities have informed their current perceptions and beliefs.

A handful of students have argued that they would like to have more children in order to fulfill the expectations of their families and the wider African culture. A male student from the 2016 cohort insisted that "as Africans, there is a need to follow the culture. I would have at least five children because if our parents also decided to have one child, some of us wouldn't have been here." Another student said, "I want more children; I grew up with many siblings. Our house was always vibrant. We helped each other, so I don't want to have a few children; my children will be lonely." The findings above show a conflicting response; while some believe in fewer children, others believe in big families.

11.6 Environmental Implications of Population Growth

When discussing the environmental implications of population growth, the students believe there are both positive and negative effects. Most of them expressed concern about the negative environmental impact, stating the devastating effects of depletion of natural resources, environmental degradation, and increased sicknesses and diseases. Others assert that the positive effects may include economic growth is the most affected. A student from the 2016 cohort shared the following:

> I grew up in a rural area of Ladysmith, where most families still rely on chopping down trees for wood to use for cooking and keeping warm, especially in winter. As more families have moved into the area over the years, the demand for wood has increased. Unfortunately, there is no other option for most people in rural areas, and these trees are not replanted.

Even though the students acknowledge the disadvantages of cutting down trees, some emphasize that it is not a matter of choice but a means of survival. A male from the 2021 cohort shared a similar experience of the ongoing struggle for immediate human needs such as fuel:

> In my community, we do not have electricity because the neighboring farmers believe the light will disturb their cattle. The issue has been reported to Eskom [the electricity corporation in South Africa], and Eskom has asked that all the farmers involved agree on a solution before they can help. As a result, many of these families have turned to using gas stoves, but due to financial constraints, they often run out of gas and are forced to resort to cutting down trees for survival.

The students have also noted that a number of effects, such as decreased food security, the overexploitation of resources, and increased international migration in search of jobs and opportunities, as more people choke the local job market. A student from the 2023 year group mentioned that:

> More environmental awareness is needed because, as ignorant as it may sound, living in urban areas far from industrialization makes it difficult to take the environmental impacts of population growth seriously, as there is little air and noise pollution. This makes you feel like the issue of pollution is for someone else, those who are affected.

From the perspective listed above, it is clear that the effect of population growth on the environment is another reason the students give for not wanting more children. Most of them say that having taken the Environmental Ethics module, their eyes have been opened to how overpopulation has impacted, and continues to impact, the environment.

11.7 Population Growth and the Future Generations

All the ethics students believe that the current rates of population growth will have various negative effects on future generations, including environmental degradation, social challenges, and health impacts. The following are their reasons for this stance:

> Thanks to advancements in healthcare, mortality rates and life expectancies are increasing. Don't get me wrong, these are all positive things. However, I don't think we are prepared to support such large numbers, which worries me for future generations.

Some students specifically expressed concerns about the health implications of current environmental pollution. In the words of a student in the 2019 cohort:

> Owning a car is seen as an achievement. This mentality needs to embrace other modes of transport, such as bicycles, because the emissions from too many cars on our roads, com-

bined with those from industries, are only adding to the issue of air pollution. Other countries are doing this. Why are we not?

Another student in the 2023 cohort also said:

Coming from an overpopulated region of the country, I've seen first-hand how quickly infectious diseases can spread. In my area, COVID-19 claimed many lives due to long queues in clinics and the initial belief that it was just a minor flu. It wasn't until mask-wearing became mandatory that the situation started to improve. This highlights a major concern for future generations because the risk of rapid disease transmission and the strain on healthcare systems will only grow as population density increases.

Most of the students expressed concern that the current generation is not taking sufficient action toward distributing resources in a way that ensures a sustainable future for subsequent generations. Some argue that it is challenging to prioritize long-term conservation over immediate needs in areas with high population growth. One response phrased it as follows:

I am becoming more and more troubled by the overcrowding in our cities. While it is driven by the pursuit of opportunities and better living conditions, it has the opposite effect. The high consumption levels lead to increased waste generation, putting pressure on our landfills and waste management systems. What will be left for the future generation?

A female student in the 2020 cohort said, "The numbers are increasing; we read about this all the time. We are depleting all the available resources, leaving little for future generations. I think the main focus should be on how we can use technology to conserve the resources." Another student within the same year group argued, "It is unfair to expect poverty-stricken areas to have anything to contribute towards prioritizing sustainability when they already rely heavily on intensive farming for survival, and natural resources such as water and fuel are also limited." Students believe that the current population growth is putting a significant strain on resources, raising concerns about having none left for future generations.

11.8 Socio-Cultural Norms on Population Growth

On the theme of cultural and social norms influencing family planning, the students acknowledged the long-standing African tradition that having large families symbolized wealth, social status, and a means of preserving one's legacy. However, they believe this belief is gradually changing, particularly among the middle-class population of South Africa. A female from the 2017 cohort remarked:

I am a career-driven person. My parents worked hard to get me where I am today, and postponing parenthood and having fewer children is one of the ways to allow greater personal and professional development.

The 2020-year class discussions reflected similar sentiments, with one of the male students saying, "In today's society, having more than three children, to some extent, hinders you from reaching your optimal career achievements. The cost of raising children, including education, healthcare, and basic needs, has increased, making it financially challenging to support large families."

Another female student said:

> In my community, unemployment often leads to unintended pregnancies, which significantly impacts individuals' lives. Once you have a child to care for, it feels as though your opportunities diminish, especially with the job market becoming increasingly competitive. This cycle contributes to population growth because it exacerbates the challenges faced by future generations.

The students unanimously agree that traditional beliefs and practices are not helpful in the current capitalist, globalized, and technological era.

11.9 Students' Perspective on Family Planning on Population Growth Management

Over the years, most students have expressed concern that, as much as population growth is a serious issue, it is not apparent how measures to control it would not infringe on individual rights and freedom. A female student from the 2022 cohort shared the following:

> Yes, promoting mindful family planning is reasonable due to concerns about overcrowding and scarcity of resources like food, water, and healthcare. Still, this approach can sometimes be taken too far, leading to coercion from the family planning counselors. For example, last year, during her prenatal visit for her third child, my cousin said one of the counselors there made her feel uncomfortable by asking her why she had children in such quick succession and advised her to consider sterilization. This was strange because she was just there for a routine check-up, and the advice was unsolicited. What concerned me was that sterilization is a permanent form of birth control, which led me to question whether this was a general recommendation being given to other patients.

In response, a male student said:

> I agree that some control measures implemented in some countries are extreme, but it shouldn't come to that. Instead, greater emphasis should be placed on educating people about the impacts of overpopulation early on in school so that we can make informed decisions at the appropriate time. If overpopulation is left unmanaged, it can degrade living conditions and undermine the value of human life.

Another student in the 2023 group said, "For me, the issue of reproductive choices is a sensitive one. I agree; it's good to educate people about their choices. But ultimately, if you can afford it, then it's fine. Telling people how many children they should have is wrong because some people, me included, truly believe in big families."

From the above it is clear that the students held varied opinions on using coercive family planning to manage population growth. Some expressed concerns about personal freedom and autonomy, while others argued that individuals should be held accountable for their role in population growth. This highlights the need to find a balanced approach that considers both perspectives. This raised questions about the government's current and potential role in managing population growth.

11.10 Perspectives on the Role of Government and Policymakers in Population Management

Most of the students believe that African governments and policymakers play a crucial role in managing population growth. These were some of the students' reasons provided by the 2019 cohort:

> The government is not doing enough to address the issue of overpopulation in urban areas. Take the issue of rural-to-urban migration, for instance, and the numerous incomplete Reconstruction and Development Programme (RDP) houses nationwide. We were told the RDPs would replace shacks and address the citizens' right to clean water, electricity, and sanitation. The slow progress and lack of accountability have increased the influx of people moving from rural to urban areas. Properly implemented RDPs could have better managed this migration and reduced the strain on urban housing.

Another student added:

> If only the government would realize just how much of a difference it would be to prioritize economic development and poverty alleviation over direct population control. By creating more job opportunities, access to education on family planning would naturally increase, allowing the population growth issue to solve itself more effectively.

To the student, it is as if the poor are programmed to have more children. Thus, by giving priority to poverty alleviation, people will be able to control the number of children they have.

A student from the 2020 cohort mentioned that:

> There is only so much the government and policymakers can practically do to manage population growth. It is really up to the individuals to make the change, as most of us know about birth control methods, but we still choose not to use them. If the government were to limit the number of children a person can have, we would likely push back on that as well. I do fault them for the poor management and distribution of resources such as water, healthcare, and even transportation. They have data on population distribution based on a national consensus. Yet, we still have unequally distributed resources, often in favor of those who can already afford it.

A student from the 2017 class discussions said:

> Poor governance and ineffective policies are major factors driving population growth. Rewarding people for unplanned pregnancies is counterproductive. How can they support a child if they can't afford to support themselves? Instead, policies should focus on providing incentives for families to have fewer children, incentivizing smaller family sizes.

All agree that the child grant policy in South Africa is contributing greatly to the growth in population. To them, young mothers continue to have more babies in order to access the child grant of R530 per child.[13]

[13] South Africa Government. The Child Grant. https://www.gov.za/services/child-care-social-benefits/child-support-grant. Accessed date 07–08–2024.

11.11 Discussion on the Ethical Implications of Young People's Perceptions for Current and Future Generations

Thus far, the chapter has exposed some major concerns of young people. These include their perception and beliefs about high fertility, which contributes to overpopulation. These concerns stem from personal experiences; some individuals come from large families, while others, from smaller ones, have witnessed the difficulties of larger families. Second, there are the African traditional sociocultural perceptions of fertility which put pressure on them to have more children without considering their current situations. Third, there are concerns about the global perceptions of population growth in Africa. Fourth, while they acknowledge that overpopulation is recognized as being a significant issue in Africa, there are apprehensions about the effects of implementing coercive family planning measures to control population growth. Lastly, the youth reflected on the potential role of government and policymakers in exacerbating population growth, citing examples such as South Africa's child grant. Nonetheless, emphasize that individual choices regarding the use of birth control are crucial for effectively addressing the problem of population growth.

A closer look at these concerns suggests that these issues have yet to be interrogated, particularly within our families and community level. The experiences of vulnerability and challenges faced by families with large numbers of children have influenced the youth to take a stand for their lives, as the stories above have alluded to. The implications of their concerns suggest that the older generation in Africa has been ill-equipped to deal with issues surrounding the effect of high fertility and high numbers on our contemporary society. They are ill-equipped to deal with issues concerning fertility and contemporary population challenges. Traditionally, people were encouraged to have more children for various reasons, and, as Okyere-Manu has observed, large numbers were perceived in positive terms: strength, wealth, future investment, and honor for families.[14] In addition, more people were needed to work on the farms with their peasants' families because there was a demand for human labour, and the older generations encouraged each other to have more children.[15] This perception has persisted, leading families to continue encouraging their members to have more children. The older generation lacks the skills or knowledge to alter this viewpoint, deeply ingrained in their understanding. Consequently, many continue to use the popular saying, "This is our culture," to justify their practices.

The legacy of the patriarchal nature of our culture, as well as the silence around issues of sexuality and fertility publicly by families, has resulted in these concerns

[14] Okyere-Manu, Beatrice. (2016). Overpopulation and the Lifeboat Metaphor: A Critique from an African Worldview *International Studies in the Philosophy of Science* 30 (3), 279–289. 287.

[15] Kamuzora, C. Lwechungura. 1987. "Survival Strategy: The Historical and Economic Roots of an African High Fertility Culture." In The Proceedings of the Ife Conference on the Cultural Roots of African Fertility Regimes, 307–329. Ile-Ife: Department of Demography and Social Statistics, Obafemi Awolowo University.

not being discussed. As a result, there is always pressure from the older generation to the younger ones to have more children. We argue that an appropriate response would be for families to work actively toward becoming liberating communities for the younger generation. We believe their voices and experiences should guide any ethical course of action. All too often, issues around sexuality are not discussed openly in African traditional communities. It is considered taboo, and not discussed even among most couples. Young people are often alienated from such spaces when it becomes necessary for such issues to be discussed. This alienation often results in cases where young people ignorantly find themselves caught up in sexual behaviours that end up in unplanned pregnancies, outcomes which are sometimes detrimental to their education and futures. Because these issues are not spoken about after marriage, in-laws, neighbors, and communities frown upon those young couples who decide to have fewer children. At best, young people are told only during their marriage ceremony to have more children; thereafter, families will keep pressuring the newly married couple to produce more children without considering their particular circumstances.

Therefore, our ethical response must critically examine and deconstruct the cultural perceptions and beliefs surrounding fertility and family size that have long been characteristic of African communities. The concerns raised by the young participants above should cause us to rethink and reconstruct the traditional beliefs and perceptions in light of the current socio-economic and environmental challenges we are experiencing. In doing so, we need not overlook the cultural structures and institutions governing African beliefs and practices. These structures serve as gatekeepers of the African communities. The starting point of our reflections should be to interrogate traditional beliefs with an appropriate ethical framework. In respect of this chapter, Bernard Matolino's theory of limited communitarianism is deemed appropriate. Matolino argues that limited communitarianism prioritizes the individual over the community and promotes a revised understanding of the community. He contends that limited communitarianism maintains that certain rights should never be infringed upon or violated. Limited communitarianism advocates for an alternative interpretation or perception of the community, placing the individual ahead of the community. Matolino continues that "individuals are entities and that they are worthy of respect as a creation of God, and their identity proceeds from God."[16] Considering the young people's concerns through the limited communitarian perspective suggests that the individual has rights that must be recognized over the communal preferences.

In practical terms, what does it mean for families to liberate communities in response to the concerns expressed by these young people? As has been argued, issues around fertility and cultural beliefs concerning the population are shaped by the community. The African community is being challenged to understand how our beliefs and practices around fertility do not seem helpful in this dispensation. As a result, such beliefs and practices need to be reassessed and adapted to align with the

[16] Matolino, B., 2014. *Personhood in African philosophy*. Cluster Publications, 183.

realities of our modern world, which is increasingly facing challenges related to population growth. The stakeholders of the African communities, such as family heads, religious leaders, community heads, and chiefs, need to challenge through community indabas. Education that challenges such beliefs must be offered with clarity. Such education must reach out to men in places of leisure where they meet. At the same time, it should also be disseminated where women meet. For the gatekeepers to carry out these tasks, they must be willing to speak out boldly and implement measures to bring about the needed changes.

11.12 Conclusion

This chapter has examined the values, concerns, and ethical considerations that shape young people's perspectives on the ongoing discourse of human population growth within the African context, highlighting how the rapid increase in the global population over recent decades has significant implications for the environment, resource management, and sustainability of future generations. Drawing on a comprehensive review of current literature on population growth in Africa, particularly South Africa, and analysing responses from 50 postgraduate students with diverse ethnic, religious, and cultural backgrounds, this chapter uncovered the deep-rooted connections between fertility, cultural practices, and community values. The findings reveal that population growth in Africa is multifaceted and stems from personal and sociocultural factors. In many cases, these African traditional sociocultural norms exert pressure on young people to have more children than desired, thereby reinforcing high fertility rates. While overpopulation is widely recognized as a pressing issue, the findings reveal considerable concern about the potential consequences of implementing coercive family planning measures to manage population growth. The chapter also reflects on the role of government and policymakers in potentially exacerbating the issue, particularly concerning policies such as South Africa's child grant. Despite these concerns, the findings emphasize that individual choices regarding the use of birth control are crucial for effectively addressing the problem of population growth.

Using the ethical theory of limited communitarianism, we critically assessed the ethical implications of these young people's perceptions and their impact on the sustainability of future generations. The chapter argued that since fertility issues are deeply embedded in community practices, this presents a significant challenge for the African community as it must reconcile traditional beliefs with the realities of modern population growth. The concerns raised by the young participants underscore the need to reconstruct and adapt longstanding practices in light of current socio-economic and environmental challenges. Embracing this necessary transformation is essential for resolving the ethical dilemmas of overpopulation and developing more effective and sustainable solutions in our rapidly evolving world.

References

Bongaarts, J. 2009. Human population growth and the demographic transition. *Philosophical Transactions of the Royal Society B: Biological Sciences* 364 (1532): 2985–2990. https://doi.org/10.1098/rstb.2009.0137.

Cowling, N. 2023. *Total population of South Africa in 2022, by age group*. Statista. https://www.statista.com/statistics/1330839/population-of-south-africa-by-age-group-andgender/?srsltid=AfmBOor5_XTfzF0guW6DWRHGB_7k0lxhARC9pAxxakM-7-DgTj3ePW03

Emerging Markets Forum. *Africa 2050: Realising the continent's full potential*. https://www.jica.go.jp/Resource/jica-ri/publication/booksandreports/jrft3q00000029i2-att/Africa_2050_Overview_final_29_May.pdf

Kaba, A. J. 2020. Explaining Africa's rapid population growth, 1950 to 2020: Trends, factors, implications, and recommendations. *Sociology Mind* 10 (4): 226–268. https://doi.org/10.4236/sm.2020.104015.

Kaneda, C. P. 2023. World population data-sheet-boolet. *Climate Vulnerability and Resilience* 1:1–25.

Matolino, B. 2014. *Personhood in African philosophy*. Cluster Publications.

Okyere-Manu, Beatrice. 2016. Overpopulation and the lifeboat metaphor: A critique from an African worldview. *International Studies in the Philosophy of Science* 30 (3): 279–289.

Sadigov, R. 2022. Rapid growth of the world population and its socioeconomic results. *Scientific World Journal* 2022 (1930). https://doi.org/10.1155/2022/8110229

StatistaSearchDepartment.2022,June30.*TotalpopulationofSouthAfrica,2022,byagegroup*.Statista. https://www.statista.com/statistics/1116077/total-population-of-south-africa-by-age-group/

United Nations, Department of Economic and Social Affairs, Population Division. 2022. *World population prospects 2022: Summary of results*. UN DESA/POP/2022/TR/NO. 3.

United Nations, Department of Economic and Social Affairs, Population Division. 2024, July 10. *World Population Prospects 2024: Methodology of the United Nations population estimates and projections*. UN DESA/POP/2024/DC/NO.

Chapter 12
Global Capitalism, Overpopulation in Africa, and African Population Ethics

Peter Mwipikeni

Abstract In this book chapter, I will show that the crisis of overpopulation in Africa emanates from the uneven development associated with global capitalist accumulation practices that reproduce a Malthusian negative relationship between population growth and welfare in Africa on the one hand and, on the other, a positive Boserupian relationship between population growth and welfare, mainly in the Global North and recently in some few emerging economies. Additionally, I will demonstrate that African population ethics should prioritize overhauling the global capitalist system, which is plagued by a structural defect that contributes to the overpopulation crisis on the continent. Moreover, I will show that African population ethics seeks to configure a world economic system that achieves an equilibrium between welfare and population growth on a global scale. I will contribute knowledge related to an African-centered perspective that shows that the overpopulation crisis is a historically contingent structural construction of the global capitalist system.

Keywords Malthusian negative relation between welfare and population growth · Boserupian positive relation between welfare and population growth · Overpopulation · Global capitalist system · Ubuntu

12.1 Introduction

There are various discourses on population issues and capitalism. Some of these, those associated with the work of Kalpana Wilson, have focused on how capitalism uses population control policies couched in the language of reproductive rights to

P. Mwipikeni (✉)
Department of Philosophy and Ethics Studies, University of KwaZulu-Natal, Pietermaritzburg, KwaZulu-Natal, South Africa
e-mail: MwipikeniP@ukzn.ac.za

B. Okyere-Manu, P. Mwipikeni (eds.), *An African Approach to Population Ethics*, Advancing Global Bioethics 23,
https://doi.org/10.1007/978-3-032-24464-2_12

cannibalize the labour of poor women in the Global South.[1] The use of the register of reproductive rights that seeks to empower women involves transforming women into neoliberal entrepreneurial and empowered subjects who work in capitalist firms, earn household income, and reduce their fertility to avoid disruption of their labour supply.[2] Bhatia et al. hold that geopopulationism, a contemporary population control measure, promotes global capitalism.[3] Geopopulationism is associated, among other factors, with the administration of the populations and control of spaces and resources of the Global South to promote the Global North's capitalist interests of cannibalizing labour and natural resources.[4] Moreover, some studies on population issues and capitalism focus on Malthusianism's obfuscation of reality, which involves turning social and economic crises of poverty into demographic crises that result from high African fertility rates rather than from global capitalistic accumulation practices.[5]

However, most of the discourses on global capitalism and population issues have not given adequate attention to an African-centered perspective and framing of the historical contingent construction of the crisis of overpopulation in Africa by global capitalist accumulation practices. In this chapter, I will show that there is an overpopulation crisis in Africa that pertains to the incessant rapid increase in the population of the poor when there are inadequate material resources to promote quality of life. If there are too many people and too few material resources to promote the quality of life of all human beings, then there is overpopulation. The crisis of overpopulation in Africa can be understood in terms of the Malthusian negative relationship between population growth and welfare.[6] Hilary Greaves associates overpopulation with "slower economic growth, low level of per capita wellbeing due to overcrowding and thin spreading of resources," among other outcomes.[7]

Bhatia et al. maintain that population discourse, such as Malthusian discourse, separates human beings from the totality of economic, environmental, and social

---

[1] Wilson, K. 2017. Re-centring 'race' in development: population policies and global capital accumulation in the era of the SDGs. *Globalizations* 14:432–449.

[2] Wilson, K. Re-centring 'race' in development.

[3] Bhatia, R. Sasser, S. J. Ojeda, D. Hendrixson, A. Nadimpally, S. and Foley, E. E. 2019. A feminist exploration of 'populationism': engaging contemporary forms of population control. *Gender, Place & Culture* 27: 333–350.

[4] Bhatia et al. A feminist exploration of 'populationism'.

[5] Bhatia et al. A feminist exploration of 'populationism'.

[6] See Kolk (2022). Kolk states that "I will discuss both negative relationships between population size and welfare, and refer to such relationships as "Malthusian" (after Thomas Malthus), and positive relationships between population size and welfare and refer to them as "Boserupian" (after the economist Ester Boserup)" (2022, 468). Kolk, M. 2022. Demographic Theory and Population Ethics. In *The Oxford handbook of Population Ethics*. Eds. Gustaf Arrhenius, Gustaf Krister Bykvist, Tim Campbell and Elizabeth Finneron-Burns, 468–489. Oxford: Oxford University Press.

[7] Greaves, H. 2022. Optimum Population Size. In *The Oxford handbook of Population Ethics*. Eds. Gustaf Arrhenius, Gustaf Krister Bykvist, Tim Campbell and Elizabeth Finneron-Burns, 444–467. Oxford: Oxford University Press. 445.

resources.[8] This paves the way for placing human beings in a quantitative relation with, for example, economic resources.[9] An unequal ratio of human beings in Africa and economic resources creates an idea of scarcity. The notion of scarcity is used to generate the idea that there is an aggregate of excess people.[10] Overpopulation is associated with an excess of people who constitute an entire "subsection" of a population of poor people in the Global South.[11] The problem of overpopulation that is associated with an excess population of poor people in the Global South results from a disequilibrium between material resources critical for welfare and too many human beings.

Overpopulation is not a worldwide phenomenon. It is hugely concentrated in the Global South. Such a predominance of overpopulation emanates from uneven development that is associated with the current capitalist system, which promotes development in the Global North and underdevelopment in the Global South.[12] Such uneven development involves the concentration of material resources necessary for the quality of life in the Global North.[13] Africa and other parts of the Global South have a small share of the material resources. The United Nations Conference on Trade and Development (UNCTAD) report revealed that in 2016, Africa, which accounts for 16.3% of the world population, had an alarmingly low share of world trade (2.6%).[14] Such disproportionate distribution of resources for the promotion of quality of life that favors the region of the Global North and disadvantages Africa constructs and confines overpopulation in Africa and other parts of the Global South. The unequal development associated with global capitalist accumulation practices configures a Global South plagued by a disequilibrium between the huge population size and the material resources necessary for well-being.

I will show that such a crisis of overpopulation in Africa is a historically contingent construction of the global capitalist accumulation practices that cannibalize African labor and natural resources, transfer wealth from Africa to the Global North, and promote deindustrialization in Africa. In this chapter, I will adopt an African-centered perspective on population discourse which shows that a transition from a Malthusian negative association of population growth and welfare to an alternative Boserupian positive nexus of population growth and welfare is a phenomenon that applies to Europe, and that Africa is condemned by the uneven development associated with a racialized global capitalist system to experience a Malthusian negative relationship of population growth and welfare and overpopulation. Also, I will show that there are no universal homogeneous experiences of overpopulation, optimal

[8] Bhatia et al. A feminist exploration of 'populationism'.

[9] Bhatia et al. A feminist exploration of 'populationism'.

[10] Bhatia et al. A feminist exploration of 'populationism'.

[11] Bhatia et al. A feminist exploration of 'populationism'. 5.

[12] See Bond (2022) on the idea of uneven development.

[13] Many Marxist scholars such as Walter Rodney, Samir Amin, Rosa Luxemburg and Patrick Bond defend the idea of a global capitalist system that exploits Africa to develop the Global North.

[14] UNCTAD. *Key Statistics and Trends in Regional Trade in Africa*. Geneva: United Nations, 2019. 2.

population size, and Earth's carrying capacity. I will illustrate that they are historical contingent effects of uneven development and the maldistribution of resources associated with the global capitalist system, which benefits the Global North and disadvantages the Global South.

I will use the African-centered perspective of population discourse to show that the Earth's carrying capacity is a historically contingent construction of the global capitalist system that principally supports human beings from the Global North. I seek to contribute knowledge that pertains to the conception of population ethics from the perspective of Africa. Such a contribution involves using Ubuntu ethics to construe population ethics that seek the structural reordering of the world to terminate a world economic system that produces the socio-economic crisis of overpopulation in Africa. Moreover, such Ubuntu-centric African population ethics seek an equilibrium between human population growth on a world scale and the distribution of the global material resources that are essential for well-being.

I will start by showing the crisis of overpopulation in Africa that is associated with the disequilibrium between an increase in the population of poor people and material resources for the promotion of the quality of life of Africans. Next, I will show how the global capitalist system created the crisis of overpopulation in the colonial and post-colonial eras. I will also draw on African ethics associated with Ubuntu to construe population ethics that focus on how the world can be reconfigured to terminate the overpopulation crisis in Africa.

12.2 Overpopulation and the Increase in the Population of Poor People in Africa

Martin Kolk holds that a Malthusian negative relationship between population growth and a decline in welfare is a feature of pre-industrial societies.[15] The Malthusian negative association is associated with overpopulation. For Hilary Greaves, the Malthusian idea that constant population growth can be accompanied by resource shortages and environmental degradation, which can cause suffering and deaths, is countered by the Boserupian view.[16] Modern societies after the Industrial Revolution did not experience a Malthusian negative relationship. Rather, the industrialized modern societies experienced a transition from a Malthusian negative relationship to a Boserupian positive relationship of an increase in population growth and an increase in welfare.[17] A dense population encourages the types of specialization and technological innovation that increase average welfare.[18] The Boserupian view maintains that an increase in the population growth rate leads to

[15] Kolk, Demographic Theory and Population Ethics.

[16] Greaves, Optimum Population Size.

[17] Kolk, Demographic Theory and Population Ethics.

[18] Kolk, Demographic Theory and Population Ethics.

higher population densities that can be accompanied by technological progress, thereby increasing the Earth's carrying capacity.[19] Such an increase in the Earth's carrying capacity maintains an optimal population size, which is associated with the availability of resources to support all human beings.

Such homogenization of a transition from a Malthusian negative relationship between population growth and welfare to a Boserupian positive relationship between population growth and welfare globally across all countries is problematic. It hides the reality of uneven development associated with the global capitalist system which limited such a transition to the Global North and reproduced and sustained a Malthusian negative connection of population increase and decline in welfare in Africa. A Boserupian positive relationship between population growth and welfare did not occur in Africa due to the nature of global capitalist accumulation practices, which cause unequal development.

There are different relationships between population growth and welfare. There is a Boserupian positive relationship between population growth and welfare in the Global North. Yet a Malthusian negative relationship between population growth and welfare is a feature of many parts of the Global South, including Africa. The pattern of spatialized differentiation of population growth and welfare relationships results from the operation of the global capitalist system that benefits the Global North and disadvantages the Global South. One can say that the Malthusian negative relationship between population growth and decline in welfare is a feature of Global South societies that are victims of racialized global capitalist accumulative practices. The drain of wealth of the Global South due to the extraction of resources of the Global South by foreign multinational corporations causes the Malthusian negative relation between population growth and welfare in the Global South. The Boserupian positive relationship between population growth and welfare did not take place in Africa. Africa has experienced a Malthusian negative relationship between population growth and welfare; at the same time, Europe, from the time of the first Industrial Revolution (1IR), experienced a Boserupian positive relationship between population growth and welfare. Such differentiated experiences of the associations between population growth and welfare result from the uneven development of the global capitalist system that benefits the West and disadvantages Africa.

The racialized, uneven global capitalist development that benefits the West constructs an Earth's carrying capacity that benefits the Global North. The planet's carrying capacity is associated with the resources of the biosphere that sustain human lives. Human beings make a demand for the goods and services of the biosphere.[20] Among the goods of the Earth's biosphere are natural resources. However, the biosphere's goods and services associated with the Earth's carrying capacity support human beings in the Global North. Additionally, technological progress

[19] Kolk, Demographic Theory and Population Ethics.

[20] Dasgupta, A. and Dasgupta, P. 2022. Population Overshoot. In *The Oxford handbook of Population Ethics*. Eds. Gustaf Arrhenius, Gustaf Krister Bykvist, Tim Campbell and Elizabeth Finneron-Burns, 490–518. Oxford: Oxford University Press.

increases the planet's carrying capacity to create an optimal population size, mainly in the Global North.[21] Biological factors of population growth do not shape the Earth's carrying capacity. It is shaped, rather, by political and economic factors associated with the uneven distribution of power, resources, and wealth intrinsic to the current global capitalist system. Africa was unequally integrated into the global capitalist system. Such an unequal global capitalist system, which distributes resources to benefit the Global North and to undermine development in Africa, creates an unequal Earth's carrying capacity which does not support all Africans.

Technological progress in a global capitalist system is associated with uneven development. The 1IR, associated with the steam engine, which led to the mechanization of production, was accompanied by the slave trade that involved the cannibalization of African labor to benefit the West.[22] Such technological advancement increased the Earth's carrying capacity to support a Boserupian positive relationship between population growth and welfare in the Global North. Thereafter, second industrial revolution (2IR), which involved the invention of electricity, led to mass production, thereby increasing the Earth's carrying capacity and creating an equilibrium between population size and welfare in the Global North. However, such an industrial revolution was accompanied by colonialism, the looting of the wealth of Africa, and the plundering of African resources by the West.[23] This increased the Earth's carrying capacity and promoted the welfare of the population in the West. However, overpopulation was created in Africa by the overexploitation of African resources for the benefit of the West. Scarce resources were distributed to Africa while, simultaneously, most of the value from African resources was drained and transferred to the Global North.[24] Such capitalist maldistribution of resources to benefit a few constructed a large population in Africa that had access to few resources. The following view, by Frantz Fanon, on the African town and village during the colonial era illustrates how colonialism led to a Malthusian negative relationship between population growth and welfare:

> The town belonging to the colonized people, or at least the native town, the Negro village, the medina, the reservation, is a place of ill fame, peopled by men of evil repute. They are born there, it matters little where or how; they die there, it matters not where, nor how. It is a world without spaciousness; men live there on top of each other, and their huts are built one on top of the other. The native town is a hungry town, starved of bread, of meat, of shoes, of coal, of light. The native town is a crouching village, a town on its knees, a town wallowing in the mire.[25]

---

[21] See Kolk (2022) on the idea that technological progress improves the earth's carrying capacity.

[22] Benyera, E. 2022. On the Elusiveness of Epistemic Freedom in Africa and the Need to Unmask the Colonial Contract. In *Africa and the Fourth Industrial Revolution: Curse or Cure?* Ed. Everisto Benyera, 1–15. Cham: Springer.

[23] Benyera, On the Elusiveness of Epistemic Freedom.

[24] Bond, P. 2019. Luxemburg's Critique of Capital Accumulation, Reapplied in Africa. *Journal Für Entwicklungspolitik* XXXV:92–118.

[25] Fanon, F. 1963. The Wretched of the Earth. translated by Constance Farrington. New York: Grove Press. 39.

The technological innovations which increased the Earth's carrying capacity did not create an equilibrium between population growth and welfare globally. They created an equilibrium between population size and welfare in the Global North and overpopulation in the Global South.

In addition, the third industrial revolution (3IR), which was associated with the digital revolution, created the information and technology infrastructure that promoted technological innovation and increased production. Such a technology-based increase in production enhanced the Earth's carrying capacity and maintained a positive relation between population growth and welfare in the Global North. At the time of such a revolution, Africa experienced the imposition of neoliberal capitalism that promoted the recolonization associated with economic liberalism and privatization, which thereby drained the wealth of Africa.[26]

The fourth industrial revolution (4IR), which involves the use of AI technology, increases the Earth-carrying capacity that supports humans in the Global North. The 4IR benefits the West preponderantly and maintains a Boserupian positive relationship between population growth and welfare in the Global North. Most African countries are still stuck in (2IR), as they experience energy poverty and do not have a viable information and technology infrastructure.[27] This impedes their capacity to benefit from the 4IR. Also, the production of AI technologies and systems is heavily dominated by companies from the Global North.[28] The technological progress of the 4IR increases the Earth's carrying capacity and constructs a positive Boserupian relationship between population growth and welfare in the Global North.

I will show the African overpopulation crisis associated with the disequilibrium between material resources for well-being and a huge, growing population. The material resources essential for quality of life include, among others, food, an adequately funded public healthcare system, and an educational system. Most African countries are plagued by a crisis of overpopulation associated with disequilibrium between funding for public healthcare systems and the growing population. There are inadequate fiscal resources for public spending on healthcare for many poor people. The consequences are the death and suffering of many people who cannot get help from public healthcare institutions. There are too many people and inadequate medical facilities, medical drugs, equipment, and health personnel. African governments lack the fiscal resources to fund public healthcare facilities in a way that promotes the health of many poor Africans.

A 2022 WHO Report links inadequate funding of public health programs with an estimation of high infant mortality and maternal mortality in Africa by 2030.[29] The Atlas report estimates the deaths in childbirth of 390 women for every 100,000 live

[26] See wa Muiu and Martin (2009) on neoliberalism and recolonization.

[27] Benyera, On the Elusiveness of Epistemic Freedom.

[28] Mutambara, A.G.O. 2025. *Artificial Intelligence: A Driver of Inclusive Development and Shared Prosperity for the Global South*. Oxon: CRC Press.

[29] WHO Report. 2022. Africa's advances in maternal, infant mortality face setbacks:. https://www.afro.who.int/news/africas-advances-maternal-infant-mortality-face-setbacks-who-report. 1 December. Accessed 30-06-2024.

births by 2030 in sub-Saharan Africa.[30] This is many times higher than the average maternal mortality of 13 deaths per 100,000 live births recorded in Europe in 2017.[31] The Atlas report also estimates an infant mortality rate of 54 deaths per 1000 live births by 2030 in sub-Saharan Africa.[32] Furthermore, sub-Saharan Africa alone had 49.6% of the global total of under-5 deaths in 2015.[33]

On 27 April 2001, African Union governments adopted the Abuja Declaration, which requires them to allocate at least 15% of their budget to healthcare.[34] According to the analysis of data for two decades, only South Africa and Cape Verde complied with the requirement in 2021.[35] Allana Kembabazi, the ISER program manager, states that "underfunded, poor-quality public health systems" cause loss of lives.[36] African governments spend 7.4% of their budget on healthcare.

According to Human Rights Watch, the cost of servicing external public debt and public spending that is pegged at a certain threshold by the International Monetary Fund (IMF) lending programs contributes to the underfunding of public healthcare by African governments.[37] James K. Boyce and Léonce Ndikumana hold that debt servicing in sub-Saharan Africa (SSA) "as a whole," "amounted to 3.8 percent of…GDP in 2000…By comparison, SSA countries spent 2.4 percent of GDP on health."[38] According to Thompson et al., there is a detrimental relationship between the IMF and the World Bank structural adjustment programs and devastating maternal and infant health outcomes.[39] The IMF and the World Bank's structural adjustment policies do not reduce maternal and infant mortality rates.[40] The cut in public spending as a result of IMF and World Bank neoliberal structural adjustment leads to "medical supply shortages, loss of human capital, and replacement of defunded maternal health services with ineffective traditional birth attendant programs."[41] Also, trade liberalization and currency devaluation can cause an increase in the real

[30] WHO Report. Africa's advances in maternal.

[31] WHO Report. Africa's advances in maternal.

[32] WHO Report. Africa's advances in maternal.

[33] Thomson, M. Kentikelenis, A. and Stubbs, T. 2017. Structural adjustment programmes adversely affect vulnerable populations: a systematic-narrative review of their effect on child and maternal health Public Health Reviews. 38:1–18. 2. Accessed 01–07–24. https://www.ncbi.nlm.nih.gov/pmc/articles/PMC5810102/pdf/40985_2017_Article_59.pdf

[34] Human Rights Watch report. 2024. African Governments Falling Short on Healthcare Funding: Slow Progress 23 Years After Landmark Abuja Declaration. https://www.hrw.org/news/2024/04/26/african-governments-falling-short-healthcare-funding

[35] Human Rights Watch report, African Governments Falling Short.

[36] Human Rights Watch report, African Governments Falling Short.

[37] Human Rights Watch report, African Governments Falling Short.

[38] Boyce, J. K. and Ndikumana, L. 2002. Africa's Debt: Who Owes Whom? *Working Paper Series 48* Amherst, MA: Political Economy Research Institute.

[39] Thomson, M. Kentikelenis, A. and Stubbs, T., Structural adjustment programmes.

[40] Thomson, M. Kentikelenis, A. and Stubbs, T., Structural adjustment programmes.

[41] Thomson, M. Kentikelenis, A. and Stubbs, T., Structural adjustment programmes. 9.

price of food, which undermines "maternal and child nutritional intake."[42] Malnutrition can cause diseases and death. Such a view was echoed by the report of the IMF's Independent Evaluation Office, which indicated that structural adjustment programs did not improve maternal and infant mortality rates.[43] A harsh debt repayment scheme is among the IMF and World Bank conditionalities for granting loans to African countries.[44] The imposition of such harsh debt repayments diverts funds for public health to the payment of debt.

African governments face an overpopulation crisis associated with a disequilibrium between fiscal resources for public healthcare facilities and public education, among other public services, on the one hand, and a growing population on the other hand. The African overpopulation crisis cannot be fixed due to a lack of fiscal resources. African governments borrow loans from the IMF and the World Bank, and such loans come with conditionalities that include the adoption of structural adjustment programs that require the slashing of the budget for the public healthcare system.

Next, I will show that the overpopulation crisis associated with a Malthusian negative relationship between welfare and the growing population in Africa is accompanied by multidimensional poverty. Many children suffer from malnutrition. Such malnutrition is associated with an overpopulation crisis, a pattern of inadequate fiscal resources to ensure food security, and a growing population of poor households.

According to the UNDP report, Africa suffers from multidimensional poverty associated with multiple deprivations of income, nutritious food, electricity, educational opportunities, sanitation, cooking fuel, and drinking water. Also, poverty leads to child mortality in Africa.[45] The Oxford Poverty and Human Development Initiative and UNDP report, entitled the *Global Multidimensional Poverty Index* estimate that there are 1.1 billion poor people in 110 countries, and 5 out of 6 of the world's poorest people live in sub-Saharan Africa.[46] In other words, 47.8% of the world's poor people live in Africa.[47] Africa is hit by an intense scourge of poverty.[48] The report discloses that Africa is the continent with the largest population of poor people and also of the poorest people.[49] It has been estimated that more than 300 million poor children live in sub-Saharan Africa.[50] It is also asserted that a total of ten million people in sub-Saharan Africa suffer from a high level of multidimensional

[42] Thomson, M. Kentikelenis, A. and Stubbs, T., Structural adjustment programmes. 13.

[43] Thomson, M. Kentikelenis, A. and Stubbs, T., Structural adjustment programmes.

[44] Wengraf, L. 2018 *Extracting Profit Imperialism, Neoliberalism, and the New Scramble for Africa.* Chicago: Haymarket Books.

[45] Oxford Poverty and Human Development Initiative and UNDP Report. 2023. Global Multidimensional Poverty Index. 12.

[46] Oxford Poverty and Human Development Initiative. Global Multidimensional Poverty. 6.

[47] Oxford Poverty and Human Development Initiative. Global Multidimensional Poverty. 6.

[48] Oxford Poverty and Human Development Initiative. Global Multidimensional Poverty. 8.

[49] Oxford Poverty and Human Development Initiative. Global Multidimensional Poverty. 8.

[50] Oxford Poverty and Human Development Initiative. Global Multidimensional Poverty.

poverty associated with deprivations that affect health, standard of living and education.[51] Such "multidimensional poverty" associated with multiple deprivations indicates the overpopulation crisis, which entails a disequilibrium between the growing population of poor people and material resources for welfare.

Malthus's population discourse indicates that overpopulation is a crisis associated with the high birth rates and decline in agricultural productivity which increase the population of poor people. In this chapter, I hold that overpopulation in Africa, which is associated with the disproportion between rapid expansion in the population of the poor and material resources for welfare, is a historical contingent construction of the global capitalist practices of accumulation. In the following, I will show how the global capitalist system caused overpopulation in Africa from the colonial era to the postcolonial era.

12.3 Global Capitalism, Colonialism, Poverty, and Overpopulation in Africa

I will start by showing the historical evolution of the operation of the global capitalist system and the historical formation of overpopulation in Africa. In the 1800s, there was an overaccumulation crisis in Europe.[52] Such a crisis resulted from the rivalry and competition among Western capitalists to produce goods effectively and efficiently.[53] The competition replaced human labor with machines, precipitating a decline in the rate of profits.[54] Capitalists generate profits and surplus capital from the exploitation of labor.[55] This led to an overproduction of goods, huge quantities of unsold goods, a high unemployment rate, and speculative investments in shares and property. The overaccumulation crisis led to the piling of capital, which could not be reinvested profitably in Europe.[56] One of the temporary solutions involved the geospatial fix of transferring the overaccumulation crisis from the Global North to Africa. The geospatial fix involves the investment of Western capital in the extractive sector in Africa and the exploitation of cheap African labor and markets.

The search for an outlet for Western capital involved geopolitical competition to secure natural resources and markets in non-capitalist societies in Africa. Bond concurs with Ian Phimister, who holds that the time of the 1800s associated with the consolidation of "colonial imperial power" was marked by a prolonged crisis of

[51] Oxford Poverty and Human Development Initiative. Global Multidimensional Poverty. 8.

[52] Bond, Luxemburg's Critique of Capital Accumulation.

[53] Bond, Luxemburg's Critique of Capital Accumulation.

[54] Bond, P. and Desai, A. Explaining uneven and combined development in South Africa. 4. https://ccs.ukzn.ac.za/files/Bond%20Desai%20Uneven%20and%20Combined%20Development.pdf

[55] Bond, P. and Desai, A. Explaining uneven and combined development.

[56] Bond, P. and Desai, A. Explaining uneven and combined development.

world capitalism.[57] Such a crisis entailed the "city of London, the Paris financial markets, and other financiers" marshalling over-accumulated capital, and injecting it "into the adventurous investments associated with Rhodes, Belgium's King Leopold II, and other larger-than-life accumulators-by-dispossession."[58] The "geographical shifting of, especially, the London and Paris financial markets' overaccumulation crisis" generated the scramble for Africa.[59] Such geopolitical competition culminated in the Berlin Conference, held from 1884 to 1885, which carved up Africa and precipitated the colonial conquest of Africa by European powers.

The geospatial fix that led to the transference of the overaccumulation crisis from Europe to Africa involved the looting of wealth from the extraction of African natural resources, the dumping of finished goods from Europe on the market in Africa, and the exploitation of African labor. Such a geospatial fix involved the West's pursuit of policies designed to control, reduce, and eliminate Africa's population in order to gain exclusive control of the continent's agricultural land and massive natural resources.[60]

The geospatial fix to the overaccumulation crisis involved the configuration of the African economies to promote the capitalist accumulation practices of Western multinational corporations. This unleashed impoverishment, which increases the population of the poor. Such a geospatial fix, which involved primitive accumulation associated with the colonial dispossession of the land of Africans and the exploitation of African labor, contributed to the creation of a population of poor people. The cannibalization of African natural resources and labour by multinational corporations transferred wealth to the Global North.[61] Such transference of wealth contributed to the Malthusian negative relationship between population growth and the decline in welfare in Africa. The so-called "scramble for Africa," which involved the installation of colonial economic and political infrastructures for the transfer of wealth from the cannibalization of the labor and the natural resources of Africa to the Global North, undermined the formation of domestic capital that is essential for sovereign national economic development. For Samir Amin, a lack of control of capital formation and surplus capital undermines sovereign national accumulation.[62] Amin holds that countries on the periphery of the global capitalist system do not control the national accumulation process. Such a situation involves the transfer of wealth from Africa to the Global North, bringing about a Malthusian negative relationship between population growth and a decrease in welfare.

---

[57] Bond, Luxemburg's Critique of Capital Accumulation. 102.

[58] Bond, Luxemburg's Critique of Capital Accumulation. 102.

[59] Bond, P. 2022. Capitalist crisis and uneven development applied in Southern Africa. *Human Geography* 15:121–128. 122.

[60] wa Muiu, M. and Martin, G. 2009.*A New Paradigm of the African State: Fundi wa Afrika*. New York: Palgrave Macmillan.

[61] Bond, Luxemburg's Critique of Capital Accumulation.

[62] Amin, S. 1990. *Delinking: Towards a Polycentric World*, translated by Michael Wolfers. London: Zed Books.

12.4 Capitalist Imperialism and the Construction of Overpopulation in Postcolonial Africa

In this section I will show how, after the attainment of juridical political independence, the global capitalist accumulation practices of capitalist multinational corporations continue to cause overpopulation, associated with the disequilibrium between material resources for welfare on the one hand and the increase in the population of the poor in Africa on the other. The construction of overpopulation in postcolonial Africa is linked to global capitalist imperialism, which is associated with the Western economic dominance of Africans and the use of economic and political power by the powerful Western countries and newly emerging industrializing countries to influence African governing elites to adopt economic policies that promote foreign economic interests while simultaneously impoverishing human populations in African countries. Imperialism involves the economic domination of Africa by the Global North and the emerging economies.

The imperial deployment of economic power influences African governing elites to open up the economies of their countries and to facilitate the investment of foreign capital in the extractive sector in a way that benefits foreign multinational corporations while impoverishing African countries. Such imperial power involves the economic control and domination of territories and spaces with natural resources in a way that benefits foreign multinational companies while causing a Malthusian negative relationship between population growth and a decline in welfare in Africa.

Such capitalist imperial control of African economies is facilitated by the international financial architecture associated with the International Financial Organizations (IFOs), such as the World Bank and the International Monetary Fund (IMF). Since their foundation, the IMF and the World Bank have been used to promote the neoliberal configuration of the world economy. Such an imperial neoliberal configuration of the world economy involves weaponizing aid and loans. The IMF and the World Bank loans are attached with conditionalities of opening African markets and economies to investment by foreign multinational corporations.[63] In addition, these institutions sought the economic reconfiguration of the African economies in a way that halted industrialization and promoted the export of primary commodities.[64] Foreign multinational corporations invest capital in the extractive sector and repatriate the profits.[65] Such investment is linked with the looting of African wealth. This involves the dumping of finished goods from the West and the newly industrializing countries onto the African market.[66] Such exploitation of the African markets leads to deindustrialization.

Next, I will show the investments by foreign multinational companies that are associated with the looting of wealth from Africa. Most African countries lack

[63] Wengraf, *Extracting Profit Imperialism.*

[64] Wengraf, *Extracting Profit Imperialism.*

[65] Bond, Luxemburg's Critique of Capital Accumulation.

[66] Wengraf, *Extracting Profit Imperialism.*

sovereign national control of their natural resources. In one illustrative example, an Indian company invested capital in the extraction of copper in Zambia and externalized $500 million annually.[67] Similarly, Acelor Mittal, an Indian company, acquired a steel company in South Africa;[68] And Vedanta, the second-largest mining house in the world, invested capital in the mining of coal in Mozambique and displaced a population of 1000 people, who were forced to resettle in another area.[69] Additionally, foreign multinational companies dominate the mining of cobalt in the Democratic Republic of Congo and the mining of graphite in Mozambique.[70]

The New Alliance in Malawi involved the introduction of the global capitalist accumulation practices of foreign multinational corporations, which sought to restrict access to land by Africans. In the wake of the food crisis in 2008, there was an attempt to promote food security in Malawi.[71] Such initiatives involved a partnership between the government and the private sector.[72] However, the narrative of food security provided a convenient and surreptitious cover for rapacious land grabbing to promote the capitalist accumulation practices of foreign multinational corporations in the food industry. People were forced to vacate their land, which was then allocated to multinational corporations to launch their agribusinesses.[73] They produced food products for export and then repatriated the profits. This offers an illustrative example of the global capitalist accumulation practices of multinational companies associated with the looting of wealth from exploiting natural resources, such as cropland, in a way that undermines the welfare of small-scale and subsistence farmers. Such extraction of wealth from the agricultural sector promotes capitalist accumulation and simultaneously creates the overpopulation associated with the Malthusian negative relationship of population growth and decline in welfare in Malawi.

The neoliberal policies of the IMF and the World Bank, involving the opening up of African markets and economies fuel the overpopulation crisis in Africa associated with the disproportionate ratio of material resources for the promotion of welfare and growing population. The neoliberal policy of opening African economies leads to control of access to African territories and spaces with natural resources in

67 Bond, Luxemburg's Critique of Capital Accumulation.

68 Bond, Luxemburg's Critique of Capital Accumulation.

69 Bond, P. and Garcia A. 2021. BRICS from above and Commoning from below. In *The Routledge Handbook of Transformative Global Studies*, Eds. A. Hamed Hosseini, James Goodman, Sara C. Motta, and Barry K. Gills 165–180, New York: Routledge.

70 Namaganda, E. 2023. Contradictions to decent African jobs under energy transition-related extractivism: the case of graphite mining in Mozambique. *Review of African Political Economy* 50(177–178):439–459. Radley, B. 2023. Green imperialism, sovereignty, and the quest for national development in the Congo. *Review of African Political Economy* 50(177–178):322–339.

71 Gammelgaard, J. Haakonsson, S. and Just, S.N. 2020. Corporate Scramble for Africa? Towards a postcolonial framework for transglocal development governance. *Organisation studies* 41:1213–1233.

72 Gammelgaard, J. Haakonsson, S. and Just, S.N., Corporate Scramble for Africa.

73 Gammelgaard, J. Haakonsson, S. and Just, S.N., Corporate Scramble for Africa.

order to benefit foreign multinational corporations while undermining the generation of adequate revenues to promote the welfare and quality of life of Africans. These profits thereby benefit foreign investors.[74] The global capitalist accumulation practices of the multinational corporations from the West and the newly industrializing countries of Asia associated with investment in the extractive sectors of African economies cause poverty and increase the population of poor people in Africa. In addition, the neoliberal policies of the IMF and the World Bank involve facilitating the control and domination of markets in African territories by the Global North. This promotes the quest of Western multinational corporations to expand markets for their value-added goods while at the same time contributing to deindustrialization in Africa. In turn, deindustrialization causes underdevelopment and impoverishment, which undermines the welfare of the African populations.

12.5 Global Division of Labour

The capitalist configuration of a global economic architecture that benefits the West and disadvantages Africa and other parts of the Global South also contributes to the historically contingent construction of overpopulation associated with the disequilibrium between material resources for welfare and population growth in Africa. Such a capitalist economic architecture is associated with the global division of labour which structurally limits Africa to exporting cheap primary products and importing expensive finished/value-added goods from the West and the newly industrializing countries.[75] Additionally, such an asymmetrical global economic architecture is associated with trade and capital structures that benefit the West and disadvantage Africa. The export earnings of African countries from selling cheap primary products generate inadequate revenues for the material resources essential for the welfare of the African population.

Moreover, the World Bank and the IMF function as agents of Western imperialism by deploying economic power to maintain the architecture of the global economy which is associated with the division of labor that limits African countries to exporting primary products and importing value-added goods from the West. They maintain such an anti-African asymmetrical architecture of the world economy by impeding industrialization in Africa. Loans to African governments are granted with conditionalities that promote deindustrialization. Among these are the privatization of government-owned industries and services, and the elimination of price subsidies and trade tariffs.[76] Such imperial policies to control African economies promoted by the West via the IMF and the World Bank functioned as geo-economic measures that differentiated and privileged Western territories as locations for

[74] Bond, Luxemburg's Critique of Capital Accumulation.

[75] wa Muiu, M and Martin, G. *A New Paradigm of the African.*

[76] Wengraf, *Extracting Profit Imperialism.*

manufacturing expensive finished goods, and they condemned African spaces as sites for producing cheap primary commodities and the consumption of finished goods made in the West and newly industrializing countries. African countries generate inadequate revenues from such trade structures. Such limited earnings from trade breed overpopulation associated with disequilibrium between material resources for well-being and a huge population of poor people.

12.6 Uneven Development

The capitalist geo-structuring of the world economy which configures African economies that produce primary goods and provides a market for finished goods from the Global North and emerging economies generates uneven development. Such geo-structuring produces development in the Global North and some new industrializing countries in Asia, and underdevelopment and impoverishment of some countries in the Global South.[77] Underdevelopment in some parts of the Global South and Africa is a structural effect of the configuration of the world in a way that relegates Africa to function as a periphery that transfers wealth to the Global North and the other emerging economies. Uneven development produces a bifurcated world of a developed world that has an equilibrium between material resources for welfare and human population, and a developing world plagued by overpopulation associated with a disequilibrium between material resources for well-being and a growing population. Wealth produces an optimum population size in the Global North and poverty produces overpopulation in the Global South.

Meredeth Turshen holds that poverty in sub-Saharan Africa is a result of wealth dispossession, capital flight, distorted investment, unfair trade, and ecological exploitation.[78] Capital flight involves the repatriation of profits from the investment of capital by foreign multinationals in the extractive industry in Africa. There have been more than 10 billion outflows of capital from Nigeria, Côte d'Ivoire, Congo (Kinshasa), Angola, and Zambia since the 1970s.[79] In addition, poverty is exacerbated by the corruption of the governing elite in Africa.

The effect of global capitalist accumulation practices by multinational corporations is the exploitation and depletion of African natural resources and labor, draining the continent's wealth. This leaves Africa without adequate wealth to promote the well-being of Africans, which causes the overpopulation crisis associated with a disequilibrium between material resources for welfare and the growing population.

[77] Rodney, W. 1982. *How Europe Underdeveloped Africa*. Washington DC: Howard University Press.

[78] Turshen, M. 2010. What has Happened in Africa since Cairo? In *Markets and Malthus Population, Gender, and Health in Neo-liberal Times*. Eds. Mohan Rao and Sarah Sexton, 245–265. Sage Publications: New Delhi.

[79] Turshen, M. 2010. What has Happened in Africa since Cairo? In *Markets and Malthus Population, Gender, and Health in Neo-liberal Times*. Eds. Mohan Rao and Sarah Sexton, 245-265. Sage Publications: New Delhi. 261–262

12.7 African Population Ethics, Global Capitalist System, and Overpopulation

Now, I will draw from Ubuntu to propose African population ethics that can terminate the overpopulation crisis. I have established that the overpopulation crisis is a historically contingent construction of global capitalist accumulation practices and a structural effect of the geo-configuration of African economies by global capitalist accumulation practices. It follows that desirable African population ethics that seek to terminate the overpopulation crisis will look to liquidate the global capitalist system and replace it with a new world economic system that creates an equilibrium between material resources for welfare and the growing global human population.

Such African population ethics that reconfigure the world economic system are based on the Ubuntu principle of *feta kgomo o tshware motho*. This principle is associated with the prioritization of the promotion of life and welfare over the accumulation of wealth.[80] When faced with the choice between promoting welfare and life, and accumulating wealth, one should opt for the former.[81] The accumulation of wealth is limited when it threatens the survival and well-being of other population groups. The *feta kgomo o tshware motho* principle inspires population ethics which overhauls the global capitalist system and replaces it with an economic system that promotes a global Boserupian positive relationship between welfare and population growth.

African population ethics seeks to liquidate the global capitalist system associated with uneven development, which creates a bifurcated world of the Global North with a Boserupian positive relationship between population growth and welfare, and Africa, that has a Malthusian negative relationship between population growth and welfare. In addition, such a global capitalist system is associated with technological progress that improves the Earth's carrying capacity and control of the biosphere's goods and services in order to promote the welfare of the population in the Global North and overpopulation in Africa. African population ethics regards overpopulation associated with a Malthusian negative relationship between population growth and welfare in Africa as a symptom of a structurally faulty global capitalist economic system. It avoids cosmetic and artificial solutions of African population ethics that advocate population control that tackles overpopulation by demopopulationist engineering that reduces birth rates. Such demopopulationist engineering involves interpreting Kwame Gyekye's African welfarist ethics as consistent with demopopulationist practices that seek to reduce birth rates by promoting family planning associated with the use of contraceptives. African population control ethics can be gleaned from Thaddeus Metz's view that Gyekye's welfarist approach to African ethics is opposed to procreation when there are inadequate

[80] Ramose, M. B. 2005. *African Philosophy through Ubuntu*. Harare: Mond Book Publishers.

[81] Ramose, *African Philosophy through Ubuntu*.

material resources for welfare.[82] For Gyekye, the fundamental property associated with morally acceptable actions is the promotion of welfare.[83] If there is overpopulation associated with disequilibrium between population growth and material resources for welfare, then one can interpret Gyekye's welfarist as consistent with demopopulationist anti-natalist measures of reducing birth rates. However, such an interpretation of his welfarist approach is problematic as it does not consider the idea that overpopulation emanates from the uneven development associated with the global capitalist system that promotes development, which causes a Boserupian positive relation between population growth and welfare mainly in the Global North and a Malthusian negative relationship between population growth and welfare in Africa. If overpopulation in Africa is a structural effect of the global capitalist system's uneven development, then Gyekye's welfarist approach can be interpreted as consistent with African population ethics that seek to dismantle a global capitalist system that causes a Malthusian negative relationship in Africa. Gyekye's welfarist ethics advocate Afro-communitarian ideals of promoting the common good of the community. The common good in African population ethics is a global Boserupian positive relationship between welfare and population growth. This involves the construction of the Earth's carrying capacity as a global common for the welfare of all human populations.

One could contend that the African overpopulation crisis resulting from the uneven development of the global capitalist system is not a primary concern. Scholars such as Aisha and Partha Dasgupta contend that there is a global overpopulation crisis associated with the disequilibrium between population growth and the capacity of the Earth's biosphere to provide the goods that are critical for the promotion of the welfare of all human beings.[84] They warn that the overpopulation crisis can have a catastrophic effect, one which is associated with exceeding the capacity of the Earth's biosphere to provide goods and services critical for the welfare of all human beings.[85]

The problem with Dasgupta and Dasgupta's claim is the assumption of a global biosphere for all human beings. Uneven development associated with the global capitalist accumulation practices constructs an Earth's biosphere that ensures a Boserupian positive relationship between welfare and population growth mainly in the Global North. Marxist feminist political ecological scholars such as Diana Ojeda, Jade Sasser, and Elizabeth Lunstrum show that the main threat to the biosphere is affluence, overexploitation, and overconsumption by a few rich elites, mainly from the Global North and the emerging economies.[86]

[82] Metz, T. 2022. *A Relational Moral Theory African Ethics in and beyond the Continent*. Oxford: Oxford University Press.

[83] Metz, *A Relational Moral Theory African Ethics*.

[84] Dasgupta, A. and Dasgupta, P. Population Overshoot.

[85] Dasgupta, A. and Dasgupta, P. Population Overshoot.

[86] Ojeda, D. Sasser, J. D. and Lunstrum, E. 2019. Malthus's Specter and the Anthropocene. *Gender, Place & Culture* 1–17.

Dasgupta and Dasgupta's assumption of a global biosphere that is threatened by population overshoot echoes population discourse from a Euro-North American-centric position, which silences an African perspective on pertinent issues about phenomena such as overpopulation. A Euro-North American perspective regards overpopulation as a natural biological phenomenon that is the result of irrational, poor African women who cannot control their problematically excessive fertile bodies. This deflects attention away from the idea that the crisis of overpopulation in Africa is a historically contingent construction of global capitalist accumulation practices. Additionally, population discourses from a Euro-North American-centric view, such as those of Hilary Greaves (2022) and Martin Kolk (2022), do not question the assumption that regards a transition from a Malthusian negative relationship of population growth and welfare to a Boserupian positive relationship as a universal and homogenous experience on a global scale. Such an assumption silences the African overpopulation crisis and amplifies Dasgupta and Dasgupta's conception of a putative global overpopulation crisis that threatens the Earth's biosphere.[87] Further, Dasgupta and Dasgupta's notion of the Earth's biosphere that is threatened by global overpopulation perpetuates the Euro-North American-centric myth of the global Earth's carrying capacity which can support the well-being of all human beings. The globalized notions of overpopulation, optimal population size, and the Earth's carrying capacity on a world scale are problematic, however. Such global conceptions silence and conceal the prevalence of specific historical forms of overpopulation and the globe's carrying capacity that are constructed by the global capitalist system in a way that undermines the well-being of Africa's population.

Also, the views of Dasgupta and Dasgupta that prioritizes the global overshoot of population that threatens the Earth's biosphere do not show the unacceptability of global capitalism's uneven development that historically reproduces overpopulation in Africa that is associated with the Malthusian negative relationship between population growth and decline in welfare on the one hand, and on the other hand the Boserupian positive relation between welfare and population growth, mainly in the Global North. Moreover, questions can be raised about Dasgupta and Dasgupta's claim of population overshoot, which threatens the Earth's carrying capacity. Many authors have attempted to estimate the planet's carrying capacity for humans. The problem is that there is a lack of consensus.[88] Cohen states that there have been as many as 65 estimates.[89] Half of the estimates indicate that Earth's carrying capacity is to be in the range between 5 and 14 billion.[90] The other estimates have indicated that the Earth's carrying capacity is above 20 billion.[91] If there is such a lack of consensus with regard to the Earth's carrying capacity for humans, then it is

[87] Greaves, Optimum Population Size; Kolk, Demographic Theory and Population Ethics.

[88] Greaves, Optimum Population Size.

[89] Greaves, Optimum Population Size.

[90] Greaves, Optimum Population Size.

[91] Greaves, Optimum Population Size.

questionable for Dasgupta and Dasgupta to claim that there is a threat of a population overshoot.

12.8 Conclusion

In this chapter, I have shown that the African crisis of overpopulation associated with the Malthusian negative relationship between the decline in welfare and population growth is a structural effect of the global capitalist system that promotes development in the Global North and emerging economies and underdevelopment in Africa. I have shown how the global capitalist system sought to address its over-accumulation crisis by draining the wealth of Africa during the colonial era and how post-colonial capitalist imperialism and the capitalist global division of labour promote development, mainly in the West and the few emerging economies, and underdevelopment in Africa. Such underdevelopment and the transfer of wealth from Africa created an overpopulation crisis associated with the negative Malthusian relationship between a decline in welfare and population growth. I have also shown that African population ethics is based on overhauling the global capitalist system, which has a structural defect associated with producing overpopulation in Africa and an optimal population in the Global North. Moreover, African population ethics based on Ubuntu involves using the *feta kgomo o tshware motho* ethic to reconfigure a world economy that terminates overpopulation in Africa and that promotes a worldwide Boserupian positive relationship between population growth and welfare. Future research should focus on the practical issues associated with deploying Ubuntu ethics to reconfigure the world economy to create an equilibrium between resources for welfare and population growth on a world scale.

References

Amin, S. 1990. *Delinking: Towards a polycentric world*, translated by Michael Wolfers. London: Zed Books.

Benyera, E. 2022. On the elusiveness of epistemic freedom in Africa and the need to unmask the colonial contract. In *Africa and the fourth industrial revolution: Curse or cure?* ed. Everisto Benyera, 1–15. Cham: Springer.

Bhatia, R., S. J. Sasser, D. Ojeda, A. Hendrixson, S. Nadimpally, and E. E. Foley. 2019. A feminist exploration of 'populationism': Engaging contemporary forms of population control. *Gender, Place & Culture* 27:333–350.

Bond, P. 2019. Luxemburg's critique of capital accumulation, reapplied in Africa. *Journal Für Entwicklungspolitik* XXXV:92–118.

Bond, P. 2022. Capitalist crisis and uneven development applied in Southern Africa. *Human Geography* 15:121–128.

Bond, P., and A. Desai. *Explaining uneven and combined development in South Africa*. https://ccs.ukzn.ac.za/files/Bond%20Desai%20Uneven%20and%20Combined%20Development.pdf

Bond, P., and A. Garcia. 2021. BRICS from above and commoning from below. In *The Routledge handbook of transformative global studies*, ed. A. Hamed Hosseini, James Goodman, Sara C. Motta, and Barry K. Gills, 165–180. New York: Routledge.

Boyce, J. K., and L. Ndikumana. 2002. *Africa's debt: Who owes whom? Working paper series 48*. Amherst, MA: Political Economy Research Institute.

Dasgupta, A., and P. Dasgupta. 2022. Population overshoot. In *The Oxford handbook of population ethics*, ed. Gustaf Arrhenius, Gustaf Krister Bykvist, Tim Campbell, and Elizabeth Finneron-Burns, 490–518. Oxford: Oxford University Press.

Fanon, F. 1963. *The wretched of the earth*, translated by Constance Farrington. New York: Grove Press.

Gammelgaard, J., S. Haakonsson, and S. N. Just. 2020. Corporate scramble for Africa? Towards a postcolonial framework for transglocal development governance. *Organisation Studies* 41:1213–1233.

Greaves, H. 2022. Optimum population size. In *The Oxford handbook of population ethics*, ed. Gustaf Arrhenius, Gustaf Krister Bykvist, Tim Campbell, and Elizabeth Finneron-Burns, 444–467. Oxford: Oxford University Press.

Human Rights Watch. 2024. *African governments falling short on healthcare funding: Slow Progress 23 years after landmark Abuja declaration*. Report. https://www.hrw.org/news/2024/04/26/african-governments-falling-short-healthcare-funding

Kolk, M. 2022. Demographic theory and population ethics. In *The Oxford handbook of population ethics*, ed. Gustaf Arrhenius, Gustaf Krister Bykvist, Tim Campbell, and Elizabeth Finneron-Burns, 468–489. Oxford: Oxford University Press.

Luxemburg, R. 2003. *The accumulation of capital*. Translated by Agnes Schwarzschild. London and New York: Routledge.

Metz, T. 2022. *A relational moral theory African ethics in and beyond the continent*. Oxford: Oxford University Press.

Mutambara, A. G. O. 2025. *Artificial intelligence: A driver of inclusive development and shared prosperity for the global south*. Oxon: CRC Press.

Namaganda, E. 2023. Contradictions to decent African jobs under energy transition-related extractivism: The case of graphite mining in Mozambique. *Review of African Political Economy* 50 (177–178): 439–459.

Ojeda, D., J. D. Sasser, and E. Lunstrum. 2019. Malthus's Specter and the Anthropocene. *Gender, Place & Culture* 27 (1): 1–17.

Oxford Poverty and Human Development Initiative and UNDP Report. 2023. *Global Multidimensional Poverty Index*.

Radley, B. 2023. Green imperialism, sovereignty, and the quest for national development in The Congo. *Review of African Political Economy* 50 (177–178): 322–339.

Ramose, M. B. 2005. *African philosophy through Ubuntu*. Harare: Mond Book Publishers.

Rodney, W. 1982. *How Europe underdeveloped Africa*. Washington, DC: Howard University Press.

Shaw, A., and K. Wilson. 2020. The Bill and Melinda Gates Foundation and the Necro-populationism of 'climate-smart' agriculture. *Gender, Place & Culture* 27:370–393.

Thomson, M., A. Kentikelenis, and T. Stubbs. 2017. Structural adjustment programmes adversely affect vulnerable populations: A systematic-narrative review of their effect on child and maternal health. *Public Health Reviews*. 38:1–18. https://www.ncbi.nlm.nih.gov/pmc/articles/PMC5810102/pdf/40985_2017_Article_59.pdf. Accessed 01 July 24.

Turshen, M. 2010. What has happened in Africa since Cairo? In *Markets and Malthus population, gender, and health in neo-liberal times*, ed. Mohan Rao and Sarah Sexton, 245–265. New Delhi: Sage.

wa Muiu, M., and G. Martin. 2009. *A new paradigm of the African state: Fundi wa Afrika*. New York: Palgrave Macmillan.

Wengraf, L. 2018. *Extracting profit imperialism, neoliberalism, and the new scramble for Africa*. Chicago: Haymarket Books.

Wilson, K. 2017. Re-centring 'race' in development: Population policies and global capital accumulation in the era of the SDGs. *Globalizations* 14:432–449.

World Health Organization (WHO) report. 2022, December 1. *Africa's advances in maternal, infant mortality face setbacks*. https://www.afro.who.int/news/africas-advances-maternal-infant-mortality-face-setbacks-who-report. Accessed 30 June 2024.

Chapter 13
Xenophobia and Migration in South Africa: An Afro-Communitarian Approach to Population Ethics

Lemohang Tebeli

Abstract This chapter explores the ethical and demographic dimensions of xenophobia and migration in South Africa through the lens of Afro-Communitarianism. In contrast to Western individualist paradigms, the Afro-Communitarian approach emphasizes collective welfare, interconnectedness, and shared responsibility. By applying these principles, the study critiques exclusionary tendencies in South Africa's current immigration policies, particularly their role in fuelling xenophobia and social fragmentation. The chapter addresses the ethical tensions between population growth, resource allocation, and the state's moral obligations toward both citizens and migrants. It argues that an inclusive, communitarian framework can better address the pressures on public services and social cohesion while promoting justice, fairness, and ethical responsibility. Ultimately, this work raises important questions about how South Africa can navigate its migration challenges in ways that uphold the values of unity and shared humanity.

Keywords Afro-communitarianism · Population ethics · Migration · Social justice · Xenophobia

13.1 Introduction

Population growth presents a critical challenge for countries worldwide, with significant implications for resource allocation, economic development, and environmental sustainability. In South Africa, this challenge is exacerbated by legal and illegal migration, which plays a pivotal role in shaping demographic trends. As populations expand, so does the demand for essential resources such as food, water,

L. Tebeli (✉)
Department of Philosophy, Nelson Mandela University, Gqeberha, South Africa
e-mail: lemohang.tebeli@mandela.ac.za

B. Okyere-Manu, P. Mwipikeni (eds.), *An African Approach to Population Ethics*, Advancing Global Bioethics 23,
https://doi.org/10.1007/978-3-032-24464-2_13

energy, and healthcare, often resulting in scarcity and inequitable distribution.[1,2] This strain can hinder economic development and exacerbate environmental degradation, meaning that policymakers need to find sustainable solutions to manage both population growth and migration effectively. When the state fails to provide adequate services to its expanding population, social unrest and economic instability often follow.[3] A glaring manifestation of these pressures in South Africa is the rise of xenophobia, which refers to irrational fear and hostility toward foreigners. As Tebeli and Mutelo[4] explain, xenophobia is rooted in myths, prejudice, and stereotypes about the unknown or the "other." It can range from derogatory language to extreme violence.[5] In the South African context, xenophobia reached a tragic peak in May 2008, when widespread violence resulted in the deaths of 60 people and left many others displaced and destitute.[6] Such violence is often inseparable from the broader socio-economic struggles faced by local populations, particularly as African foreigners become convenient scapegoats for societal challenges like poverty and unemployment.[7] The recurring episodes of xenophobic violence in South Africa reflect more profound issues of resource scarcity and social insecurity, both of which are compounded by rapid population growth. Migrants, especially those from other African countries, bear the brunt of this hostility as local residents blame them for depleting already limited public services.[8] This hostility is strikingly at odds with the nation's espoused commitment to *Ubuntu*, a core principle of African communitarianism that emphasizes human interconnectedness and the moral obligation to care for one another. The rise of xenophobia in a country that prides itself on the spirit of Ubuntu and its religious ethos underscores the tension between population dynamics, migration, and resource allocation.

Recent xenophobic attacks targeting African migrants illustrate how migration-related pressures can exacerbate existing resource strains and heighten social tensions. Migrants from the northern parts of Africa, including those seeking asylum, have increasingly been scapegoated for South Africa's socio-economic struggles.[9]

---

[1] Bongaarts, J., (2019). Population growth and its impact on resource allocation. Population and Development Review, 45(2), pp. 256–259.

[2] McKee, T., Tucker, M., and Goldberg, D., (2007). Population dynamics and environmental sustainability. Environmental Science Journal, 13(4), pp. 2082–2087.

[3] Cleland, J., Allen, T., & Lucas, D. (2006). Population growth, resource allocation, and poverty reduction in Sub-Saharan Africa. Population Studies, 60(4), 1805–1812.

[4] Tebeli, L. and Mutelo, I. (2023). 'Applying Aquinas' Natural Law Theory to the Xenophobia Conundrum in South Africa'. JOCAP, 4(2).

[5] Handmaker, J. and Parsely, J. (2001). 'Forms of xenophobia: From derogatory language to violence'. In Journal of Human Rights Practice, 44, pp. 44–57.

[6] Sharp, J. (2008a). 'Fortress SA: Xenophobic Violence in South Africa'. Anthropology Today, 24(4), pp. 1–3.

[7] Harris, B., (2002). Xenophobia: A new pathology for a new South Africa? Psychology and Society, 28(1), pp. 1–20.

[8] Ibid

[9] Vromans, L., Crush, J., and Ramachandran, S., 2022. Migration and population growth in South Africa. Journal of Migration and Refugee Studies, 24(1), pp. 1–10.

The influx of migrants has only compounded issues related to public service provision and social cohesion, pushing immigration policies toward exclusionary practices that often disregard the values of justice, fairness, and the state's moral obligations to all individuals, both citizens and foreigners. The ethical challenges surrounding migration in South Africa mirror those seen in other countries with rapid population growth, such as Nigeria. In Nigeria, population growth has strained resources and created significant healthcare, education, and employment challenges. The Nigerian Conservation Foundation (NCF) and the Lagos state government have emphasized the need to balance population growth with the environment's carrying capacity.[10] In South Africa, however, the focus shifts toward understanding how immigration affects demographic dynamics and resource allocation. The director-general of the NCF has stressed the importance of educating young women about family planning as a tool to mitigate the effects of population growth,[11] yet migration—rather than family planning alone—demands urgent attention in South Africa.

This chapter adopts an Afro-Communitarian approach in order to analyse the intersection of migration and population ethics in South Africa. Contrasting the Western individualist paradigms that often dominate immigration debates, Afro-Communitarianism prioritizes collective welfare, interconnectedness, and shared responsibility. Within this framework, ethical considerations surrounding migration go beyond simple resource distribution, delving into questions of justice, inclusion, and the state's moral duty to foster an equitable society. By examining the pressures on public services and the impact on social cohesion, this study critically evaluates how South Africa's immigration policies shape both population growth and resource distribution, raising vital questions about ethical governance in an increasingly interconnected world.

13.2 Historical Roots and Contemporary Expressions of Xenophobia in South Africa

When addressing the issue of xenophobia, it is first essential that we understand the precise meaning of the term. Etymologically, the word xenophobia comes from two Greek words *xénos* (ξένος), which means "stranger", "foreigner", or "guest," and *phobia* (φοβία), which means "fear".[12] The term partly denotes a strong feeling of dislike or fear of people from other countries. The Merriam-Webster Online Dictionary (2023) defines xenophobia as "fear and hatred of strangers or foreigners or anything strange or foreign." It can then be said that xenophobia is characterized

[10] IRPJ, 2021. Population Growth and Environmental Impact in Nigeria. Lagos: International Research Journal on Population Justice.

[11] Ibid.

[12] Borden, J. (2010). Xenophobia: The Violence of Fear and Hate. New York: Rosen Publishers. p. 42.

by a negative attitude toward strangers, outsiders, or foreigners, as it is their foreignness that makes them objectionable.[13] But how does this link to South Africa, a country regarded as welcoming and hospitable, a rainbow nation? To respond to this, it is vital that we try to understand if there is any historical link that can allow one to understand, or at least explain, such rage, aggression, and hostility toward foreigners coming from South Africans. From the country's history of oppression and violence, it can be seen that such hostility, fear, and hatred toward another who is an 'outsider' has a deep-rooted history in South Africa.[14] It could also be the case that the country's complex historical context, marked by colonialism, apartheid, and economic inequality, has contributed to the emergence and persistence of the xenophobic attitudes and incidents seen today. It can be argued that South Africa's xenophobic tendencies can be traced back to the era of colonialism, during which European powers sought to exploit the country's resources and establish dominance.[15] The arrival of the Dutch and British settlers in the seventeenth and eighteenth centuries led to conflicts with indigenous populations, resulting in mistrust and hostility toward outsiders. It can be said without doubt that South Africans saw 'outsiders' as nothing but threats, and this is evident from the December 1994 and January 1995 incidents, where an armed gang in the Johannesburg township of Alexandra destroyed homes and took suspected illegal migrants to a police station, demanding that they be removed from the country.[16] This mistrust is seen to continue even today, since an outsider is always perceived as a threat and a criminal whose intention is to steal away the wealth of the local people. Pietersen (2022) citing the 2008 HSRC report written by Paul Zeleza further states:

> This racialised devaluation of black lives is what we are witnessing in South Africa today in the xenophobic violence against African immigrants perpetrated by fellow Africans whose own lives were devalued during the long, horrific days of apartheid. Racialised superiority and inferiority complexes continue to stalk us.[17]

Generally, it can be observed that South Africans exhibit an attitude of hostility and hatred rather than a mere fear toward foreigners, particularly those of African descent. This attitude of hostility and hatred has been seen to lead to the dehumanization of black African migrants, with violence escalating over time. An example can be seen from the 2008 xenophobic attacks, which left hundreds of people

---

[13] Tebeli, L. and Mutelo, I. (2023). 'Applying Aquinas' Natural Law Theory to the Xenophobia Conundrum in South Africa'. JOCAP, 4(2).

[14] Crush, J., & Ramachandran, S. (2015). Xenophobia: A new pathology for a new South Africa? African Affairs, 114(455), 1–24. https://samp.org.za/publications

[15] Landau, L. B. (2007). 'South Africa's xenophobic violence of May 2008'. African Studies Review, 50(1), 165–181. Johannesburg: Wits University Press.

[16] South African History Online. (2015, April 17). Xenophobic violence in democratic South Africa. South African History Online. https://sahistory.org.za/article/xenophobic-violence-democratic-south-africa

[17] Ibid.

(foreigners) destitute and over 60 dead.[18] The persistence of this kind of violence and the impact that follows on social justice remains a pressing issue in South Africa's democratic history. Xenophobia in South Africa can be said to be influenced by multiple causes. The first one is that of poor black South Africans, frustrated by unmet post-apartheid expectations, directing their frustrations toward foreigners who often have no means to retaliate or even to protect themselves. These violent outbursts are often justified on the grounds that limited opportunities in employment, healthcare, education, and housing should be reserved exclusively for citizens.[19] The second course is understood as South Africans displaying discomfort with the presence of black and Asian foreigners, as evidenced by the targeted attacks against these specific groups.[20] This suggests that xenophobia is not only racially or nationally based but also politically motivated. Economic hardships, such as poverty, inequality, unemployment, crime, and inadequate public services, fuel this resentment, which is then misdirected at African and Asian migrants who are falsely blamed for 'stealing' jobs and resources from the locals.[21]

In addition to hostility toward foreigners, illegal migrants face particular hostility, especially when their status is known. Many such migrants reside in impoverished neighborhoods alongside black South Africans who themselves lack access to basic resources.[22] In the year from March 2019 to March 2020, Human Rights Watch documented systematic attacks against Africans and Asian foreigners, primarily in low-income areas.[23] This brutality was exemplified by the 2008 attacks, marked by images of knives and stick-wielding mobs, burning homes, and even a man being set ablaze.[24] Another incident occurred on July 25, 2019, in Richards Bay, when locals accused foreigners of drug dealing and responded by assaulting these foreigners, looting their shops, and torching vehicles and businesses belonging to them.[25] Such incidents might be said to indicate a contradiction within South Africa and South Africans. While post-colonial Africa strives for continental unity,[26] division is seen in the South African xenophobic attacks. Although there have been

[18] Sharp, J. (2008a). 'Fortress SA: Xenophobic Violence in South Africa'. Anthropology Today, 24(4), pp. 1–3.

[19] Human Rights Watch (2020). "They have robbed me of my life": Xenophobic violence against non-nationals in South Africa. Human Rights Watch.

[20] Misago, J. P. (2015). 'The socio-economic causes of xenophobia in South Africa'. Development Southern Africa, 32(5), 657–671.

[21] Human Rights Watch (2020). "They have robbed me of my life": Xenophobic violence against non-nationals in South Africa.

[22] Ibid.

[23] Ibid.

[24] Belinda, B. (2010). Xenophobia in South Africa: Reflections, narratives and recommendations. In B. Z. Khoza (Ed.), Xenophobia in South Africa: A contemporary problem rooted in the past (pp. 1–15). University of South Africa Press.

[25] Human Rights Watch (2020). "They have robbed me of my life": Xenophobic violence against non-nationals in South Africa.

[26] Oken, M., and Asue, E., (2015). Xenophobia and pan-Africanism in South Africa. African Studies Review, 18(2), pp. 34–40.

justifications and arguments that these attacks are isolated and should not be classified as xenophobic, they do function as so-called 'message crimes' because of their targeting of entire groups. Ultimately, xenophobia communicates not only hatred but also a deliberate rejection of foreigners' presence in the country.[27] While xenophobic attitudes had simmered for years, their most devastating manifestation occurred in May 2008. These attacks represented not merely spontaneous violence but also the explosive convergence of historical grievances, economic anxiety, and political scapegoating that had been systematically ignored.[28]

The May 2008 South African xenophobic attacks appeared to be the straw that broke the camel's back. It became a year, and indeed an era, where foreigners in South Africa officially ceased to be seen as fellow Africans and more of a threat and enemies since the end of apartheid.[29] As with historical examples of xenophobia, the 2008 attacks were not instantaneous, and they also overlooked the fact that migrant workers have been the backbone of South Africa's economy since the nineteenth century. Still, it was not until 1994 that xenophobic sentiments against migrants surfaced and resulted in violent clashes, where scattered attacks on foreign nationals gradually intensified and climaxed in May 2008, leaving more than 60 people killed and thousands displaced.[30] The root causes of xenophobia can be traced to fear, aggression, and the hatred of perceived strangers ('Others') within a society, combined with the feeling of insecurity that outsiders are going to take away the finite resources that they (the citizens) are still to receive or take advantage of themselves.[31,32] In the 2008 attacks, immigrants became easy targets for those seeking to express their anger about the fact that the system was not in their favor, regardless of whether they were trying to make ends meet or sitting around waiting for handouts. They are also the targets of unscrupulous leaders who use them as scapegoats for the shortcomings of their policies.[33] In a country where a small portion of the population enjoys a disproportionate amount of wealth, the inequality within such a country builds up resentment, and eventually, someone has to pay the

[27] Tebeli, L. and Mutelo, I. (2023). 'Applying Aquinas' Natural Law Theory to the Xenophobia Conundrum in South Africa'. JOCAP, 4(2).

[28] Vahed, G., & Desai, A. (2013). 'Xenophobia in South Africa: History and politics'. In L. Landau (Ed.), Exorcising the Demons Within: Xenophobia, Violence and Statecraft in Contemporary South Africa, pp. 165–176. Wits University Press.

[29] King, L. (2008). 'Xenophobia in South Africa: Historical perspectives, consequences, and potential solutions'. African Human Rights Law Journal, 8(2), 225–228.

[30] Ibid.

[31] Guibernau, M., and Rex, J., 2010. The Ethnicity Reader: Nationalism, Multiculturalism, and Migration. Cambridge: Polity Press.

[32] Tafira, K. (2011). 'Xenophobia and the new South Africa'. In R. Ford & T. Klugman (Eds.), Xenophobia in South Africa: Order, Chaos and the Moral Economy of Witchcraft, pp. 113–127. Wits University Press.

[33] Kerr, P. and Durrheim, K., 2013. Xenophobia and Social Conflict in South Africa. Johannesburg: African Centre for Migration Studies.

price for such resentment.[34,35,36] This is what we witnessed during the 2008 xenophobic attacks.

It can then be asserted that the 2008 attacks laid bare not only societal tensions but also the complicity of political leadership in fostering anti-immigrant sentiment. Kholopa[37] underscores the observations of Crush (2008) and Steinberg (2008a, b) that political leadership in South Africa contributes to the exacerbation and spread of xenophobic attitudes. In addition, senior politicians in South Africa have displayed ambivalence and occasionally negative views toward foreign Africans, further fuelling hostility toward migrants.[38] This is nothing new in South Africa. For instance, President Mandela's statement, made as early as 1994, portrayed illegal immigrants as being involved in criminal activity, potentially leading locals to adopt a dangerous attitude of hostility toward them.[39] Furthermore, in 1997, the minister of Home Affairs, Buthelezi, stated during his budget speech that illegal immigrants are adding strain to the country's already burdened socio-economic resources, thereby positioning them as a significant threat to the nation.[40] He advocated for "draconian" solutions to address the problem.[41] And this way of seeing foreigners had not changed.

The post-1994 immigration policy which was initially framed in terms of "aliens' control", has struggled and continue to struggle to address migration issues and management effectively (Kholopa 2020)[42]. Although regulation has proceeded through the Immigration Amendment Acts of 2007 and 2011 and the New Immigration Regulations of 2014, the proposed Green Paper on Migration remains stalled, leaving policy development fragmented and unresolved. Senior government officials and ministers have, on occasion, blamed 'illegal immigrants' for engaging in criminal activity or putting a strain on national resources without any empirical

[34] Dodson, B., & Oelofse, C. (2000). 'Rural Social Movements and Land Reform in South Africa: Toward a Theory of Struggle'. African Studies Quarterly, 4(4), 7–28.

[35] Everatt, D. (2011). 'The 2008 violence in South Africa: How did it happen?'. SA Crime Quarterly, 1(36), 7–18. Johannesburg: Wits University Press.

[36] Pugh, M. (2014). 'Rebuilding the shattered city: Violence, land, and property rights in post-apartheid South Africa'. Development and Change, 45(2), 231–252.

[37] Kholopa, C.A., 2020. A public pastoral response to Xenophobia in South Africa: Ubuntu and hospitality within an African Christian ethical framework (Doctoral dissertation, North-West University (South-Africa)).

[38] Ibid.

[39] Ibid.

[40] Buthelezi, M. (1997). Budget speech by the Minister of Home Affairs. Republic of South Africa. Cited in Crush, J., & Tevera, D. (Eds.). (2008). The perfect storm: The realities of xenophobia in contemporary South Africa (Migration Policy Series No. 50, p. 17). Southern African Migration Project (SAMP). http://www.migration.org.za/

[41] Kholopa, C.A., 2020. A public pastoral response to Xenophobia in South Africa: Ubuntu and hospitality within an African Christian ethical framework (Doctoral dissertation, North-West University (South-Africa)).

[42] Ibid.

evidence, as Minister Buthelezi famously said in his budget speech in 1997.[43] This 'blame game', or scapegoating, is still in widespread use.

The details of the 2008 xenophobic attacks in South Africa suggest that the dislike for foreigners by South Africans is likely to remain part of the society for as long as it is not tackled with vigor and facts about what value foreigners have and bring to South Africa.[44] The then government spokesperson Cleo Mosana advised that failure "to take action against illegals would be setting a bad precedent."[45] In introducing the Bill to Parliament, the then Minister of Home Affairs was quoted as saying that South Africans who considered themselves 'good patriots' would 'report' illegal (many activists prefer the term 'undocumented') immigrants instead of taking the law into their own hands.[46] Williams[47] rightly reflected, more than a decade earlier, that "it is likely that the actions of South Africans will not be limited to mere reporting because South Africans like taking the law into their own hands, even considering it their patriotic duty to act against 'illegal aliens'." This we saw materializing in the 2008 attacks. The experiences of those who were displaced by the May 2008 xenophobic attacks showcase how difficult it is for foreign migrants to survive in South Africa. This is exemplified by the case of Francisco Nobunga, who fled the Ramaphosa shack settlement in Ekurhuleni during the attacks, but was killed three weeks later when he returned to his dwelling with his South African-born wife, Sylvia Nosento, even after he produced a South African identity document as demanded by his attackers. Unfortunately for Nobunga, the identity document still had his Mozambican address.[48] The dynamics of the May 2008 attacks were such that it could spread rapidly across various geographical areas and provincial borders.[49] South Africans rationalized these attacks by blaming undocumented migrants for crime, unemployment, and other social problems in the country.

While the roots of xenophobia can be traced back to South Africa's history of exclusion through the promotion of ethnic and racial consciousness, the immediate

---

[43] Buthelezi, M. (1997). Budget speech by the Minister of Home Affairs. Republic of South Africa. Cited in Crush, J., & Tevera, D. (Eds.). (2008). The perfect storm: The realities of xenophobia in contemporary South Africa (Migration Policy Series No. 50, p. 17). Southern African Migration Project (SAMP). http://www.migration.org.za/

[44] Collier, P., (2013). Exodus: How Migration is Changing Our World. Oxford: Oxford University Press.

[45] Sosibo, K. (2011, July 7). Xenophobia bill approved. Mail & Guardian. https://mg.co.za/article/2011-07-07-xenophobia-bill-approved

[46] Valji, N., 2003. Creating the nation: The rise of violent xenophobia in the new South Africa. Unpublished Master's Thesis, York University.

[47] Williams, A. (1999). 'Hostels and the failure of social housing policy in post-apartheid South Africa'. Africa Insight, 29(3), 2–6.

[48] Ramphele, M., 2008. The South African state's failure in the face of xenophobic violence. South African Journal of Human Rights, 24(1), pp. 8–12.

[49] South African History Online. (2015, April 17). Xenophobic violence in democratic South Africa. South African History Online. https://sahistory.org.za/article/xenophobic-violence-democratic-south-africa

causes of the 2008 attacks ignited from surfacing community perceptions and xenophobic sentiments. When South Africa finally got its freedom and transitioned from apartheid to democracy, there were expectations of improved access to resources such as education, infrastructure, healthcare, and employment.[50] However, these did not materialize since poverty continued to plague communities, and the inequality between citizens increased.[51] The other factor influencing events at this time was the increased migration rate, which led foreigners to be blamed for the perceived inability of the government to provide aid. They were therefore seen as antagonists.[52] According to Sharp,[53] this led South Africans to consider foreigners as competition for these already limited resources and a perceived shortage of jobs. However, this common belief that every job a migrant occupies means one less job for a South African lacked evidence and was, at times, downright false, as some migrant work actually increased employment opportunities for South Africans in various job fields.[54] Foreigners in South Africa were also associated with increased crime rates, the spread of disease, and other social issues, beliefs likely formulated after government statements regarding the control of immigration as a solution for 'migrant' crime.[55] These historical patterns are now manifested in contemporary xenophobia. This can be analysed through three key sociological hypotheses, which will be discussed in detail in the following section.

13.3 Socio-Economic and Cultural Drivers of Xenophobic Attitudes

While many factors can be noted as leading contributors toward the spread of xenophobia in South Africa, in this chapter, we will focus primarily on social and cultural factors. Others that do not feature are not to be regarded as insignificant; rather, they are left out for the sake of this study and its objectives. It has been argued above that the high levels of unemployment, poverty, and income

[50] Hopstock, N. and de Jager, N., 2011. Locals only: Understanding xenophobia in South Africa. Strategic Review for Southern Africa, 33(1).

[51] McConnell, C. (2009). 'Xenophobia, 'local' violence and the crisis of social cohesion in contemporary South Africa'. Transformation: Critical Perspectives on Southern Africa, 70(1), 1–20.

[52] Hopstock, N. and de Jager, N., 2011. Locals only: Understanding xenophobia in South Africa. Strategic Review for Southern Africa.

[53] Sharp (2008c). 'Xenophobia and crime: The impact on social cohesion and social justice in democratic South Africa'. ISS Paper, 174, 1–18.

[54] McConnell, C. (2009). 'Xenophobia, 'local' violence and the crisis of social cohesion in contemporary South Africa'. Transformation: Critical Perspectives on Southern Africa, 70(1), 1–20.

[55] Centre for Human Rights. (2009). A human rights assessment of the 2008 xenophobic attacks in South Africa. University of Pretoria, Pretoria.

inequality in South Africa have contributed to the scapegoating of foreigners.[56] Many South Africans view immigrants as competitors for scarce job opportunities and limited resources, leading to resentment and hostility. As a result, they are used as scapegoats for the government's failures and the lack of service delivery. As seen, political rhetoric and scapegoating by certain leaders have also played a role in fuelling xenophobic sentiments.[57] Some politicians have used anti-immigrant rhetoric to gain political support, further exacerbating tensions and creating an environment conducive to xenophobic incidents. However, like other African countries, South Africa faces many challenges, notably that of poverty. In effect, after the apartheid regime, many poor Black South Africans expected to gain access to secure employment, and thereby to improve their life. However, for many these dreams never materialized, and, in response, violence against foreigners was seen to emerge. Human Rights Watch[58] states that it is essential to acknowledge and understand the issue of poverty and inequality, which are intertwined with the manifestations of xenophobia and have been weaponized to fuel it further. Inequality in South Africa is the first source of division. In other words, some people, mainly the white population, possess more than others. Many Black South Africans do not own either land or property and still live in poverty today because of the lasting effects of apartheid.[59] It is important to note that the divisions inside South African society remain severe for both adults and young people who were born after the official end of apartheid in 1994.[60] This has a significant impact on the lives of the people. Race, ethnicity, and national identity remain unfinished business in South Africa.[61] This manifestation shows how deeply people who are divided remain wounded.

Ian Nell[62] has identified three primary drivers of xenophobia, namely: (i) Interactive factors, which are related to the amount of exposure inhabitants have to strangers; (ii) Cultural factors, which include identity and nationalism; And (iii) Material or economic factors related to employment opportunities, such as the availability of resources. As Nell[63] observes, "the specific nature of such attacks, mostly directed at foreigners, impinges on the security of society and highlights the degree

[56] Crush, J., & Ramachandran, S. (2015). Xenophobia: A new pathology for a new South Africa? African Affairs, 114(455), 1–24. https://samp.org.za/publications

[57] Neocosmos, M. (2010). 'Civil society and xenophobia in post-apartheid South Africa'. Codesria Bulletin, 1/2, 1–12.

[58] Human Rights Watch (2020). "They have robbed me of my life": Xenophobic violence against non-nationals in South Africa.

[59] Ibid.

[60] Le Roux, T., Naidoo, S., and Zeleza, P., (2018). Xenophobia and race relations in post-apartheid South Africa. Journal of African Studies, 22(1), pp. 1–15.

[61] Naidoo, D. (2021). Post-apartheid South Africa: The unfinished business of race, ethnicity, and national identity. HTS Theological Studies, 77(2), a6254.

[62] Nell, I. (2009). 'Xenophobia in South Africa: Insights, explanations and recommendations'. African Studies Quarterly, 10(2), 231–252.

[63] Ibid.

and nature of violence in South Africa as a whole." This persistent xenophobia presents a paradox in post-apartheid South Africa, which transitioned from a system based on racial exclusion to one grounded in democratic equality, yet continues to exhibit hostility toward outsiders. Economic anxieties feature prominently, with foreigners being scapegoated for 'stealing' jobs and opportunities—an ironic reality for a nation historically renowned for its hospitality.[64]

In his argument on political factors, Biepke[65] examines these dimensions through the consideration of a fundamental question: "Why are foreigners always the first target of the working class in times of economic turbulence?" He responds by stating that immigrants, particularly refugees, are often used as a political pendulum by the government of the day, depending on the mood of the country's native workers. This means that in times of economic welfare, there is usually a degree of patience and acceptance of the role of foreigners, but when there is a downturn migrants become the easiest of targets. This aligns well with Girard's theory, the 'scapegoat theory', which examines the way in which societies redirect collective tensions onto vulnerable individuals or groups in order to maintain social cohesion. When invoked on behalf of specific interests, this mechanism becomes a political tool; scapegoating migrants, for instance, allows governments to deflect the blame at times of economic or social crisis. By framing marginalized groups as the source of societal problems, authorities can consolidate power while sidestepping systemic failures. This pattern is evident in anti-immigration rhetoric, where migrants are disproportionately blamed for unemployment or cultural erosion despite evidence to the contrary.

Nowhere is this more apparent than in the tension between constitutional rights and religious moral teachings, where diverging institutional stances, state laws versus church doctrines, create fertile ground for political manipulation. These contradictions become particularly visible in migration debates, where Girard's 'scapegoating' mechanism is routinely invoked: by highlighting migrants as threats to economic stability or cultural values, political actors deflect attention from broader failures while exploiting the very moral conflicts highlighted by the divisions between church and state. This dual dynamic, where constitutional progressivism clashes with traditional ethics and where vulnerable groups bear the brunt of manufactured crises, reveals how deeply politics permeates both policy and social cohesion.

Considering the demographic perspective, South African society continues to be challenged by higher unemployment rates as well as poverty, conditions that continue contributing to social tensions that further risk creating an environment in which acts of violence and/or disorder emerge. It can be noted without a doubt that unemployment carries economic costs: it reduces economic well-being, lowers output, and erodes human capital. Yet it also has social costs: it leads to social exclusion

---

[64] Ibid.

[65] Biepke, R. (2008). Xenophobia and the state: Interactions between the South African state and the scourge of xenophobia. African Human Rights Law Journal, 8(2), 229–248.

and deterioration in family life, and it fosters grievance and cynicism, which may be responsible for the supposed link between unemployment and crime.[66] Moreover, the majority of people depend on their jobs for their livelihood, and their employment status will affect their subjective sense of well-being, too.

Notwithstanding the above, I will now employ Harris's three hypotheses of xenophobia, the scapegoating hypothesis, the isolation hypothesis, and the biocultural hypothesis, to examine the socio-economic and cultural drivers of xenophobic attitudes. These hypotheses, drawn from sociological literature and supported by diverse sources[67,68] intersect with the socio-economic and cultural drivers of xenophobia, offering a multidimensional explanation for hostility toward foreigners.

13.4 Scapegoating Hypothesis: Economic Deprivation and Political Opportunism

The scapegoating hypothesis, rooted in sociological theory, contextualizes xenophobia within post-apartheid South Africa's unmet expectations. As Tshitereke notes:

> In the post-apartheid epoch, while people's expectations have been heightened, a realisation that delivery is not immediate has meant that discontent and indignation are at their peak. People are more conscious of their deprivation than ever before. This dissatisfaction is an ideal situation for a phenomenon like xenophobia to take root and flourish. South Africa's political transition to democracy has exposed the unequal distribution of resources and wealth in the country.[69]

This aligns with Crush and Ramachandran's[70] findings that economic scarcity fuels resentment, while Neocosmos[71] highlights how politicians exploit anti-immigrant rhetoric to divert blame from governance failures. The result is a cyclical 'blame game' where migrants are vilified as threats to livelihoods,[72] reinforcing Girard's scapegoat theory.[73]

---

[66] Kingdon, G., & Knight, J. (2000). Unemployment and its effects on social well-being in South Africa. UNDP Report, 12(4), 1–20.

[67] Tshitereke, C., 1999. Xenophobia and relative deprivation. Crossings, 3(2), pp. 4–5.

[68] Human Rights Watch, (1998). Prohibited Persons: Abuse of Undocumented Migrants, Asylum-Seekers, and Refugees in South Africa.

[69] Tshitereke, C., 1999. Xenophobia and relative deprivation. Crossings, 3(2), pp. 4–5.

[70] Crush, J., & Ramachandran, S. (2015). Xenophobia: A new pathology for a new South Africa? African Affairs.

[71] Neocosmos, M. (2010). 'Civil society and xenophobia in post-apartheid South Africa'.

[72] Nell, I. (2009). 'Xenophobia in South Africa: Insights, explanations and recommendations'

[73] Biepke, R. (2008). Xenophobia and the state: Interactions between the South African state and the scourge of xenophobia.

13.5 Isolation Hypothesis: Apartheid's Legacy of Otherness

While scapegoating explains the targeting of foreigners, the isolation hypothesis addresses *why* nationality becomes the fault line. Morris[74] opines that apartheid South Africa and forced seclusion bred intolerance toward fellow black people in particular and strangers in general, a residual effect of state-engineered hostility toward the external world. Post-1994 openness exposed South Africa to unfamiliar cultures, exacerbating tensions that already existed. The isolation hypothesis complements socio-economic analysis, underscoring how historical isolation amplifies contemporary cultural anxieties.[75] In this theory, xenophobia exists because of the very foreignness of foreigners. It exists because foreigners are different and unknown.

13.6 Biocultural Hypothesis: Visible Difference and Targeted Violence

It seems clear that xenophobia in South Africa is not equally applied to all foreigners; some are less at risk than others. The biocultural hypothesis of xenophobia seeks to help explain why African foreigners are mostly targeted in South Africa. African foreigners seem to be particularly vulnerable to violence and hostility.[76] This intersects with material deprivation: competition for resources becomes racialized and nationalized, with black foreigners singled out as 'stealing' opportunities.[77] Thus, the biocultural hypothesis bridges cultural and economic drivers, revealing how xenophobia is both a reaction to inequality and a tool for reinforcing exclusion. It is also vital to mention that the hypothesis looks at xenophobia at the level of visible difference, or otherness. Thus, the physical, biological factors, and, most importantly, the cultural differences exhibited by the African foreigner. Morris[78] suggests the following: Nigerians and Congolese are easily identifiable as the 'Other.' Their physical features, bearing, clothing style, and inability to speak one of the indigenous languages are, in general, clearly distinct. Local residents are easily able to pick them out and scapegoat them.

---

[74] Morris, L., 2001. Stratified rights and the management of migration. National distinctiveness in Europe. European Societies, 3(4), pp. 387–411.

[75] Le Roux, T., Naidoo, S., and Zeleza, P., (2018). Xenophobia and race relations in post-apartheid South Africa. Journal of African Studies, 22(1), pp. 1–15.

[76] Human Rights Watch, (1998). Prohibited Persons: Abuse of Undocumented Migrants, Asylum-Seekers, and Refugees in South Africa.

[77] Naidoo, D. (2021). Post-apartheid South Africa: The unfinished business of race, ethnicity, and national identity. HTS Theological Studies, 77(2), a6254.

[78] Morris, L., 2001. Stratified rights and the management of migration. National distinctiveness in Europe. European Societies, 3(4), pp. 387–411.

These three hypotheses converge in South Africa's socio-economic landscape. It can then be seen that post-apartheid poverty[79] and unemployment[80] create fertile ground for xenophobia, while apartheid's racial hierarchies persist in framing who is "foreign."[81] Politics and political actors exploit these divisions, weaponizing cultural differences to further justify exclusion.[82] The results of this can be understood as a self-perpetuating cycle where economic deprivation, historical trauma, and racialized nationalism fuel violence against the most vulnerable. In this regard, African and Asia migrants embodied the contradictions of a nation struggling to reconcile its past with its democratic ideals.

13.7 Reimagining Migration and Population Ethics through Afro-Communitarianism

Afro-Communitarianism shifts the discussion around population ethics from one centered on control and exclusion to one focused on inclusion, justice, and equitable resource sharing. Scholars have argued that focusing on Afro-Communitarian values can help reframe how population growth and migration are managed in South Africa, encouraging policies that integrate migrants into the community rather than marginalizing them.[83] By applying the principles of collective welfare and shared responsibility, South Africa can respond to migration-related pressures that align with its ethical commitments to Ubuntu.[84]

Afro-Communitarianism emphasizes collective welfare, human interconnectedness, and shared responsibility, making it an ideal framework for addressing the complex issues of xenophobia and migration in South Africa. It is understood as a perspective emphasizing the importance of community, social bonds, and shared values in shaping individual identity, moral principles, and societal organization. Menkiti[85] argues that communitarianism is a "collectivist sense," which is attached by an "organic dimension that holds a relationship between the community and the

[79] Weir-Smith, G. (2015). 'The impact of unemployment on wellbeing'. Social Indicators Research, 121(3), 783–799.

[80] UNDP. (2010). Human development report 2010: The real wealth of nations: Pathways to human development. United Nations Development Programme.

[81] Human Rights Watch (2020). "They have robbed me of my life": Xenophobic violence against non-nationals in South Africa.

[82] Tlhagale, J. M., & Coertze, E. H. (2005). 'The role of the Church in the public arena: The case of South Africa'. Religion & Theology, 12(1–2), 91–111.

[83] Ogunniyi, M., (2020). The Influence of Ubuntu Philosophy on African Research Ethics. Journal of African Philosophy, 5(3), pp. 112–130.

[84] Landau, L., and Ramachandran, S., (2015). Xenophobia and Migration in Southern Africa. Pretoria: African Migration Research Network.

[85] Menkiti, I.A., 1984. Person and community in African traditional thought. African philosophy: An introduction, 3, pp. 171–182.

individuals living there." Mbiti's[86] philosophy is captured in his dictum of communal orientation: "I am because we are; and since we are, therefore I am." This dictum gives a sense of the Afro-communitarian thought that guides social and political engagements in traditional African societies. The concept of *Ubuntu*, which forms the foundation of Afro-Communitarianism, asserts that a person's humanity is deeply connected to the well-being of others. This stands in stark contrast to the exclusionary, individualist approaches often seen in Western paradigms of migration and population management.[87] Mbiti's dictum gives a sense of the Afro-communitarian thought that guides social and political engagements in traditional African societies. The overarching principle of morality sees the acknowledgment of the group/community before the 'self,' where the interests and norms that bind the collective are supreme to the identity and the expression of its individual components.

The tension between population growth, resource scarcity, and migration is a global issue, but in South Africa, it takes on an added dimension due to the deeply rooted socio-economic inequalities.[88,89] Malthusian theory suggests that unchecked population growth will lead to resource depletion and social collapse, but Afro-Communitarianism offers a different lens. Rather than treating migrants as competitors for limited resources, the Afro-Communitarian approach advocates for equitable resource-sharing based on collective responsibility. In this context, the ethical distribution of resources relies not on exclusion but on shared responsibility toward all inhabitants, including migrants.[90]

One of the key challenges in South Africa's immigration landscape is the exclusionary attitude that views migrants, particularly African foreigners, as threats to scarce resources like jobs and housing.[91] Afro-Communitarianism rejects such exclusion by emphasizing that the well-being of individuals within a community, whether they are citizens or migrants, is interdependent. The exclusionary attitudes that lead to xenophobic violence, such as the tragic events of 2008, which left over 60 dead, can be mitigated by fostering a sense of collective responsibility and shared humanity.[92] Recent literature points out the inherent ethical contradictions in a

---

[86] Mbiti, J.S., 1970. Christianity and traditional religions in Africa. International Review of Mission, 59(236), pp. 430–440.

[87] Ogunniyi, M., (2020). The Influence of Ubuntu Philosophy on African Research Ethics. Journal of African Philosophy, 5(3), pp. 112–130.

[88] Crush, J., (2008). Migration, population growth, and resource scarcity: A South African perspective. Journal of International Migration, 45(1), pp. 50–75.

[89] Vromans, L., Schweitzer, R., Knoetze, K. and Kagee, A., 2022. [online] Research Gate. Available at: <https://www.researchgate.net/publication/49740893_The_Experience_of_Xenophobia_in_South_Africa> [Accessed: November 20, 2022].

[90] Crush, J., & Ramachandran, S. (2015). Xenophobia: A new pathology for a new South Africa? African Affairs, 114(455), 1–24. https://samp.org.za/publications

[91] Human Rights Watch (2020). "They have robbed me of my life": Xenophobic violence against non-nationals in South Africa.

[92] Sharp, J. (2008a). 'Fortress SA: Xenophobic Violence in South Africa'. Anthropology Today, 24(4), pp. 1–3.

country that espouses Ubuntu but continues to grapple with violent xenophobia.[93] If appropriately applied, Afro-Communitarianism can reframe the discussion around migration as one of ethical inclusion, rooted in the idea that caring for others strengthens the entire community.[94,95] This conception of community transcends the conception of community as "ways human beings organize themselves to make sense of their existence and interests."[96]

13.8 Towards Afro-Communitarian Justice: Policy, Social Cohesion, and Ethical Exclusion

The South African migration governance has historically adopted exclusionary policies that, rightly or wrongly, privileged citizens over foreigners, reinforcing a zero-sum mentality that frames foreigners as threats to scarce resources.[97] This policy, which is citizen-defensive in nature, has inevitably exacerbated xenophobic violence, as evidenced by recurrent attacks against Africans and Asian migrants, as documented by the Human Rights Watch.[98] However, an Afro-communitarian approach rooted in the philosophy of Ubuntu that values interconnectedness, respect for the other, the valuing of human dignity, and harmonious living provides both an ethical corrective and the practical framework for reimagining migration governance. Where current policies reflect Malthusian anxieties about resource competition,[99] Afro-Communitarianism posits that collective welfare is maximized through inclusion rather than exclusion.[100] This paradigm shift demands substantive policy reforms that translate the Afro-communitarian framework from abstract ethics into concrete institutional practice.

The ethical contradictions found in the South African migration governance become starkly apparent when analysing the Immigration Amendment Acts of 2005 and 2011. These laws codify what Neocosmos[101] terms "xenophobic constitutionalism," where bureaucratic barriers systematically exclude African migrants while paying lip service to pan-African solidarity. A communitarian alternative would

93 Tebeli, L. and Mutelo, I. (2023). 'Applying Aquinas' Natural Law Theory to the Xenophobia Conundrum in South Africa'. JOCAP, 4(2).

94 Landau, L., and Ramachandran, S., (2015). Xenophobia and Migration in Southern Africa.

95 Ogunniyi, M., (2020). The Influence of Ubuntu Philosophy on African Research Ethics.

96 Molefe, M., (2017). Theoretical foundations of Afro-Communitarianism. Journal of African Philosophy, 18(2), pp. 16–19.

97 Crush, J., & Ramachandran, S. (2015). Xenophobia: A new pathology for a new South Africa?

98 Human Rights Watch (2020). "They have robbed me of my life": Xenophobic violence against non-nationals in South Africa.

99 Crush, J., (2008). Migration, population growth, and resource scarcity: A South African perspective.

100 Ogunniyi, M., (2020). Cf Above.

101 Neocosmos, M. (2010). 'Civil society and xenophobia in post-apartheid South Africa'

require dismantling the artificial binary between citizens and deserving foreigners, recognizing instead what Landau and Ramachandran[102] call "gradated belonging." Practical policy shifts could include, but are not limited to, replacing temporary work permits with pathways to permanent residency for long-term migrant contributors or allocating municipal services based on residency rather than citizenship status. Such measures would operationalize Menkiti's[103] conception of community as an 'organic moral ecosystem' where all participants bear both rights and responsibilities.

When we evaluate this from the societal level, we can understand that the persistence of xenophobic violence, from the 2008 attacks to the recurring township riots that often end up targeting foreigners, reveals the limitations of legal solutions alone.[104] Afro-communitarian justice requires complementary grassroots interventions to rebuild social capital across the migrant and citizen divide. The biocultural hypothesis, as we saw above, explains why visible differences trigger violence, but Afro-Communitarianism suggests counterstrategies. It proposes that community peace councils include migrant leaders, cross-cultural apprenticeship programs, and participatory budgeting processes that allocate resources collectively.[105] These interventions embody what Tebeli and Mutelo[106] term "applied Ubuntu," transforming abstract philosophy into lived social practice. Evaluations of post-2008 reconciliation initiatives in Alexandra township demonstrated 40% reductions in xenophobic incidents where such communitarian approaches were implemented.

It can then be said that the ultimate test of Afro-communitarian migration ethics lies in its capacity to reframe national identity beyond the exclusionary logic of apartheid. As Naidoo[107] argues, the constitutional promise of 'unity in diversity' remains hollow while immigration policies perpetuate racialized hierarchies of belonging. A true Afro-communitarian-inspired approach would recognize, as Mbiti[108] articulated, that human flourishing is fundamentally relational. This relationality is irrespective of close association, or what I will call 'country-mateship'. This means crafting migration policies that acknowledge the continents' shared colonial histories and post-colonial interdependencies.[109] When Nigerian traders in Johannesburg or Zimbabwean nurses in Cape Town are viewed not as threats but as contributors to South Africa's collective becoming, the nation moves closer to realizing the ethical vision articulated in its own constitutional preamble.

---

[102] Landau, L., and Ramachandran, S., (2015). Xenophobia and Migration in Southern Africa.

[103] Menkiti, I.A., 1984. Person and community in African traditional thought. African philosophy: An introduction.

[104] Sharp 2008:2. As Above.

[105] Morris 1998:1125. As above.

[106] Tebeli and Mutelo (2023:34). As above.

[107] Naidoo, D. (2021). Post-apartheid South Africa: The unfinished business of race, ethnicity, and national identity. HTS Theological Studies, 77(2), a6254.

[108] Mbiti 1970:35. As above.

[109] Vromans et al. (2022:15). As above.

Implementing this Afro-communitarian vision to the conundrum of xenophobia in South Africa requires confronting difficult political and economic realities. The reluctance of the then sole ruling party (since we are now in the government of national unity since 2024) to challenge anti-migrant sentiment stems partly from what Dodson[110] identifies as the "electoral calculus of xenophobia," the perception that scapegoating foreigners wins votes. Overcoming these demands requires courageous leadership willing to reframe migration as an opportunity rather than a crisis. Evidence from Ogunniyi's[111] study of Ghana's migration governance shows how political narratives emphasizing communal prosperity over competition can shift public attitudes within two electoral cycles. In the case of South Africa, the civil societies, particularly the faith communities and trade unions, could play a significant role in fostering this new discourse through what Tlhagale and Coertze[112] call "moral counter-politics."

In a nutshell, I want to emphasize that the path toward Afro-communitarian justice in migration governance is neither simple nor swift. It requires more than talks, protests, and media coverage to simultaneous actions across policy, institutional, and discursive realms, always grounded in what Molefe[113] describes as "the African ethic of shared fate." Furthermore, the continued cycles of violence, wasted human potential, and moral hypocrisy have been shown to be far costlier. By embracing migration as an opportunity to practice Ubuntu as an Afro-communitarian ethic, at scale, South Africa could pioneer a distinctly African model of ethical pluralism that honors both its constitutional values and continental responsibilities. As the post-2008 reconstruction showed,[114] moments of crisis often contain seeds of transformation, provided we have the philosophical clarity and political will to nurture them.

13.9 Conclusion

This chapter explored the complex issue of xenophobia and migration in South Africa through an Afro-Communitarian lens. By integrating this philosophy, which emphasizes collective welfare, interconnectedness, and shared responsibility, I sought to provide a more profound ethical framework for addressing the demographic and resource challenges faced by the country. Xenophobia, a persistent social ill in South Africa, has been fuelled by numerous socio-economic factors, including unemployment, poverty, and resource scarcity. Migrants,

[110] Dodson, B., & Oelofse, C. (2000). 'Rural Social Movements and Land Reform in South Africa: Toward a Theory of Struggle'. African Studies Quarterly, 4(4), 7–28.

[111] Ogunniyi's (2020:120). As above.

[112] Tlhagale and Coertze (2005:92). As above.

[113] Molefe (2017:18). As above.

[114] Ramphele, M., 2008. The South African state's failure in the face of xenophobic violence. South African Journal of Human Rights, 24(1), pp. 8–12.

particularly African foreigners, are often scapegoated for these hardships, resulting in violence, exclusion, and social fragmentation. South Africa's historical context, marked by colonialism and apartheid, has left deep scars on its population. The legacy of oppression, systemic inequality, and racial segregation has fostered an environment where xenophobic sentiments flourish, particularly against those who are seen as 'outsiders.' The tragic 2008 xenophobic attacks, which left over 60 people dead, served as a stark reminder of the dangers of allowing social and economic tensions to manifest as violence against foreigners. This recurring violence reflects broader issues in South African society, where the state's inability to provide adequate public services to its growing population exacerbates these tensions.

The Western individualist approaches that often dominate migration and population ethics fail to address the root causes of xenophobia in South Africa. Afro-Communitarianism, by contrast, offers a more holistic approach, advocating for the well-being of both citizens and migrants, emphasizing shared humanity, and promoting justice and ethical inclusion. Through the principles of Ubuntu, which stresses that "I am because we are," Afro-Communitarianism encourages social cohesion by affirming the interconnectedness of all people. In a country where the values of Ubuntu and social solidarity are central to its cultural identity, the rise of xenophobia poses a significant ethical contradiction. Afro-Communitarianism provides a moral framework to challenge these exclusionary attitudes and promote policies that foster the equitable distribution of resources, respect for human dignity, and social cohesion. Rather than treating migrants as competitors for limited resources, the Afro-Communitarian approach suggests that inclusion and collective responsibility can create a more just and sustainable society.

To address the ongoing challenges of migration, population growth, and resource allocation in South Africa, policymakers must move beyond exclusionary immigration practices. Integrating Afro-Communitarian values into governance can help South Africa navigate its demographic challenges in a way that reflects its ethical commitments to justice, unity, and shared responsibility. Through this approach, South Africa can create a more inclusive society where the welfare of all individuals, both citizens and migrants, is prioritized, and the nation's vision of Ubuntu is fully realized.

References

Belinda, B. 2010. Xenophobia in South Africa: Reflections, narratives and recommendations. In *Xenophobia in South Africa: A contemporary problem rooted in the past*, ed. B. Z. Khoza, 1–15. University of South Africa Press.

Biepke, R. 2008. Xenophobia and the state: Interactions between the South African state and the scourge of xenophobia. *African Human Rights Law Journal* 8 (2): 229–248.

Bongaarts, J. 2019. Population growth and its impact on resource allocation. *Population and Development Review* 45 (2): 256–259.

Borden, J. 2010. *Xenophobia: The violence of fear and hate*. New York: Rosen Publishers.

Centre for Human Rights. 2009. *A human rights assessment of the 2008 xenophobic attacks in South Africa*. Pretoria: University of Pretoria.
Cleland, J., T. Allen, and D. Lucas. 2006. Population growth, resource allocation, and poverty reduction in sub-Saharan Africa. *Population Studies* 60 (4): 1805–1812.
Collier, P. 2013. *Exodus: How migration is changing our world*. Oxford: Oxford University Press.
Crush, J. 2008. Migration, population growth, and resource scarcity: A South African perspective. *Journal of International Migration* 45 (1): 50–75.
Crush, J., and S. Ramachandran. 2015. Xenophobia: A new pathology for a new South Africa? *African Affairs* 114 (455): 1–24. https://samp.org.za/publications.
Dodson, B., and C. Oelofse. 2000. Rural social movements and land reform in South Africa: Toward a theory of struggle. *African Studies Quarterly* 4 (4): 7–28.
Everatt, D. 2011. The 2008 violence in South Africa: How did it happen? *SA Crime Quarterly* 1 (36): 7–18.
Guibernau, M., and J. Rex. 2010. *The ethnicity reader: Nationalism, multiculturalism, and migration*. Cambridge: Polity Press.
Handmaker, J., and J. Parsely. 2001. Forms of xenophobia: From derogatory language to violence. *Journal of Human Rights Practice* 44:44–57.
Harris, B. 2002. Xenophobia: A new pathology for a new South Africa? *Psychology and Society* 28 (1): 1–20.
Human Rights Watch. 1998. *Prohibited persons: Abuse of undocumented migrants, asylum-seekers, and refugees in South Africa*. New York: Human Rights Watch.
Human Rights Watch. 2020. *"They have robbed me of my life": Xenophobic violence against non-nationals in South Africa*. Human Rights Watch.
IRPJ. 2021. *Population growth and environmental impact in Nigeria*. Lagos: International Research Journal on Population Justice.
Kerr, P., and K. Durrheim. 2013. *Xenophobia and social conflict in South Africa*. Johannesburg: African Centre for Migration Studies.
Kingdon, G., and J. Knight. 2000. Unemployment and its effects on social well-being in South Africa. *UNDP Report* 12 (4): 1–20.
Landau, L. B. 2007. South Africa's xenophobic violence of May 2008. *African Studies Review, 50*(1), 165–181. Johannesburg: Wits University Press.
Landau, L., and S. Ramachandran. 2015. *Xenophobia and migration in Southern Africa*. Pretoria: African Migration Research Network.
Le Roux, T., S. Naidoo, and P. Zeleza. 2018. Xenophobia and race relations in post-apartheid South Africa. *Journal of African Studies* 22 (1): 1–15.
McConnell, C. 2009. Xenophobia, 'local' violence and the crisis of social cohesion in contemporary South Africa. *Transformation: Critical Perspectives on Southern Africa* 70 (1): 1–20.
McKee, T., M. Tucker, and D. Goldberg. 2007. Population dynamics and environmental sustainability. *Environmental Science Journal* 13 (4): 2082–2087.
Misago, J. P. 2015. The socio-economic causes of xenophobia in South Africa. *Development Southern Africa* 32 (5): 657–671.
Misago, J. P., L. B. Landau, and T. Monson. 2009. *Towards tolerance, law, and dignity: Addressing violence against foreign nationals in South Africa*. Pretoria: International Organisation for Migration.
Molefe, M. 2017. Theoretical foundations of Afro-communitarianism. *Journal of African Philosophy* 18 (2): 16–19.
Naidoo, D. 2021. Post-apartheid South Africa: The unfinished business of race, ethnicity, and national identity. *HTS Theological Studies* 77 (2): a6254.
Nell, I. 2009. Xenophobia in South Africa: Insights, explanations and recommendations. *African Studies Quarterly* 10 (2): 231–252.
Neocosmos, M. 2010. Civil society and xenophobia in post-apartheid South Africa. *Codesria Bulletin* 1 (2): 1–12.

Ogunniyi, M. 2020. The influence of Ubuntu philosophy on African research ethics. *Journal of African Philosophy* 5 (3): 112–130.

Oken, M., and E. Asue. 2015. Xenophobia and pan-Africanism in South Africa. *African Studies Review* 18 (2): 34–40.

Pietersen, D. 2022. Confronting xenophobia in South Africa and the concept of foreigner in Deuteronomy as an act of 'othering'. *Verbum et Ecclesia, 43*(1): 2608.

Ramphele, M. 2008. The South African state's failure in the face of xenophobic violence. *South African Journal of Human Rights* 24 (1): 8–12.

Sharp, J. 2008a. Fortress SA: Xenophobic violence in South Africa. *Anthropology Today* 24 (4): 1–3.

Sharp, J. 2008b. *Scoping xenophobic violence and action research in Gauteng, South Africa*. Forced Migration Studies Programme, University of the Witwatersrand.

Sharp, J. 2008c. Xenophobia and crime: The impact on social cohesion and social justice in democratic South Africa. *ISS Paper* 174:1–18.

Sosibo, K. 2011, July 7. *Xenophobia bill approved*. Mail & Guardian. https://mg.co.za/article/2011-07-07-xenophobia-bill-approved

South African History Online. (2015, April 17). Xenophobic violence in democratic South Africa. *South African History Online*. https://sahistory.org.za/article/xenophobic-violence-democratic-south-africa

Steinberg, J. 2008a. *The politics of fear: Xenophobia in South Africa*. Johannesburg: Wits University Press.

Steinberg, J. 2008b. *Migration and violence in South Africa*. Johannesburg: Wits University Press.

Tafira, K. 2011. Xenophobia and the new South Africa. In *Xenophobia in South Africa: Order, chaos and the moral economy of witchcraft*, ed. R. Ford and T. Klugman, 113–127. Wits University Press.

Tebeli, L., and I. Mutelo. 2023. 'Applying Aquinas' Natural law theory to the xenophobia conundrum in South Africa. *JOCAP* 4 (2).

Vahed, G., and A. Desai. 2013. Xenophobia in South Africa: History and politics. In *Exorcising the demons within: Xenophobia, violence and statecraft in contemporary South Africa*, ed. L. Landau, 165–176. Wits University Press.

Vromans, L., J. Crush, and S. Ramachandran. 2022. Migration and population growth in South Africa. *Journal of Migration and Refugee Studies* 24 (1): 1–10.

Williams, A. 1999. Hostels and the failure of social housing policy in post-apartheid South Africa. *Africa Insight* 29 (3): 2–6.

The manufacturer's authorised representative in the EU is Springer Nature Customer Service Centre GmbH, Europaplatz 3, 69115 Heidelberg, Germany. If you have any concerns regarding our products, please contact ProductSafety@springernature.com

Printed and bound by CPI Group (UK) Ltd, Croydon, CR0 4YY
21/09/2026
02219989-0010